Joining Technology of γ-TiAl Alloys

Joining Technology of γ-TiAl Alloys

Sónia Simões
Filomena Viana
Manuel F. Vieira
CEMUC
Faculdade de Engenharia da Universidade do Porto (FEUP)
Universidade do Porto
Porto
Portugal

CRC Press is an imprint of the
Taylor & Francis Group, an **informa** business
A SCIENCE PUBLISHERS BOOK

Cover illustrations reproduced with kind permission of the first author of the book, Sónia Simões

CRC Press
Taylor & Francis Group
6000 Broken Sound Parkway NW, Suite 300
Boca Raton, FL 33487-2742

First issued in paperback 2020

ISBN-13: 978-1-4987-3874-3 (hbk)
ISBN-13: 978-0-367-78209-2 (pbk)

Library of Congress Cataloging-in-Publication Data

Names: Simões, Sónia. | Viana, Filomena. | Vieira, Manuel F.
Title: Joining technology of gamma TiAl alloys / Sónia Simões, Filomena Viana, Manuel F. Vieira, CEMUC, Faculdade de Engenharia da Univeridade do Porto (FEUP), Universidade do Porto, Porto, Portugal.
Other titles: Joining technology of gamma titanium-aluminum alloys
Description: Boca Raton, FL : CRC Press, 2017. | "A Science Publishers book." | Includes bibliographical references and index.
Identifiers: LCCN 2017008753| ISBN 9781498738743 (hardback : alk. paper) | ISBN 9781498738750 (e-book : alk. paper)
Subjects: LCSH: Titanium-aluminum alloys. | Joints (Engineering)
Classification: LCC TN693.T5 S537 2017 | DDC 669/.7322--dc23
LC record available at https://lccn.loc.gov/2017008753

Visit the Taylor & Francis Web site at
http://www.taylorandfrancis.com

and the CRC Press Web site at
http://www.crcpress.com

Preface

γ-TiAl alloys are a novel class of advanced structural materials which enlarge the possible choices of materials for high temperature applications, especially when the weight reduction of the components is critical for structural applications. The development and processing of high temperature materials is crucial to technological advances in engineering areas where materials have to meet extreme demands. γ-TiAl alloys have a series of properties that make them attractive to the aerospace and automotive industries and are being used to produce components for the new generation of aircraft engines and automotive turbochargers.

The joining of components is an essential part of a man-made product manufacturing. The joining processes often contribute significantly to the cost of the final product and to the production difficulties. γ-TiAl alloys can be successfully joined by processes such as friction welding, brazing, diffusion bonding and transient liquid phase bonding. These processes do not require the melting of the base material and therefore the joints do not present high residual stress, so that the formation of some deleterious intermetallics can be prevented. However, very demanding bonding conditions are needed to produce a sound interface with the desired mechanical properties. Even when the bulk materials remain at temperatures significantly below their melting point (such as in soldering, brazing and diffusion bonding), the required processing temperatures often induce phase transformations, structural modifications and localised stresses that impair the properties of the joints.

The joining of γ-TiAl alloys, either to themselves or to dissimilar materials, requires the application of new strategies in order to obtain good quality joints in appropriate processing conditions. This book describes the current state of knowledge regarding the bonding of these advanced materials, with emphasis on the techniques used by the authors, who for more than a decade, have undertaken research into the joining of these materials by solid-state diffusion bonding and brazing, together with teams from the Universities of Coimbra and Minho.

This book has been organised into eight chapters—three introductory chapters covering the applications, structure and conventional processes for joining γ-TiAl alloys, followed by five chapters focusing on the use of reactive multilayers to bond γ-TiAl alloys to themselves and to other materials. This book is intended for students, materials scientists, engineers, and technicians who desire to familiarise themselves with the joining of γ-TiAl alloys and with the new developments in this field.

Sónia Simões
Filomena Viana
Manuel F. Vieira

Acknowledgements

This book is the result of a collaborative research work and fruitful discussions with many colleagues working in the field of joining of advanced materials. Though it would be impossible to thank all those who have directly or indirectly contributed to this project, a few specific references will be made to the following contributors.

We would like to acknowledge the invaluable cooperation of the team from the University of Coimbra (CEMUC), Prof. Dr. Maria Teresa Vieira and Dr. Ana Sofia Ramos, in all aspects related to the reactive multilayer systems and to the team from the University of Minho, Dr. Ana Maria Pinto and Dr. Aníbal Guedes, in the brazing of γ-TiAl alloys. The authors are also very grateful to Dr. Mustafa Koçak for his scientific and technical support and to the facilities provided at the Helmholtz-Zentrum Geesthacht.

We would like to acknowledge the financial support of the Fundação para a Ciência e Tecnologia-FCT and the Faculty of Engineering, University of Porto-FEUP.

Finally, we would like to thank our colleagues, co-workers and students at the Department of Metallurgical and Materials Engineering, University of Porto, for their encouragement and suggestions.

Contents

List of Abbreviations

AFM	Atomic Force Microscopy
ARB	Accumulative Roll Bonding
BM	Base Material
BSE	Electron Backscattered
DSC	Differential Scanning Calorimetry
E	Young's Modulus
EBSD	Electron Backscatter Diffraction
EDS	Energy-Dispersive X-Ray Spectroscopy
EELS	Electron Energy-Loss Spectroscopy
FFT	Fast Fourier Transform
FSW	Friction Stir Welding
HAZ	Heat-Affected Zone
HIP	Hot Isostatic Pressing
HRTEM	High Resolution Transmission Electron Microscopy
OM	Optical Microscopy
PBHT	Post-Bond Heat Treatment
PM	Powder Metallurgy
PVD	Physical Vapor Deposition
RT	Room Temperature
SAED	Selected Area Electron Diffraction
SEBM	Selective Electron Beam Melting
SEM	Scanning Electron Microscopy
SLM	Selective Laser Melting
SPF	Superplastic Forming
SPF/DB	Superplastic Forming and Diffusion Bonding
SPS	Spark Plasma Sintering
STEM	Scanning Transmission Electron Microscopy
TEM	Transmission Electron Microscopy
TLP	Transient Liquid Phase
TMAZ	Thermo-Mechanically Affected Zone
UTS	Ultimate Tensile Strength
VAR	Vacuum Arc Melting
WZ	Weld Zone
XRD	X-Ray Diffraction
YS	Yield Strength

List of Figures

List of Tables

CHAPTER 1

γ-TiAl Alloys

1.1 Introduction

The development and processing of high temperature materials is crucial for technological advances in engineering areas where materials have to meet extreme conditions. The increasing demand for reduction in weight for a better fuel economy and an operating efficiency requires the development of weight-saving structural materials with high service temperatures. The properties of titanium aluminide alloys make them good candidates for meeting these requirements (Dimiduk 1999, Knippscheer and Frommeyer 1999, Appel and Oehring 2003, Peters et al. 2003, Lütjering and Williams 2007, Appel et al. 2011, Kothari et al. 2012, Bewlay et al. 2013). In applications for the aerospace and automobile industries, they have become an alternative to heat-resistant steels and nickel-based superalloys. For instance, gamma titanium aluminide (γ-TiAl) alloys are used to produce low pressure turbine blades for the new generation of aircraft engines (GEnx-1B and GEnx-2B produced by General Electric Aviation, CFM LEAP by CFM International and PurePower®PW1000G by Pratt and Whitney) and turbine wheels of the new generation of turbochargers (engine Twin Turbo V-6, which equips the 2016 Cadillac GT6).

The main attractive properties of these alloys are low density, high-specific strength, high-specific stiffness, good oxidation resistance and good creep properties at high temperatures. Besides, the burn resistance of the γ-TiAl alloys makes it possible for the substitution of some expensive alloys used in aircraft engines. The specific stiffness of these alloys is also very attractive, since it is 50 per cent greater than the conventional structural materials used in these applications (Appel and Oehring 2003, Lütjering and Williams 2007, Appel et al. 2011, Kothari et al. 2012, Froes 2015).

The high melting point and ordered crystalline structure of these intermetallic alloys are the result of the strong bond between the atoms. This limits the dislocation movement and atomic diffusion, improving the high temperature properties, but dramatically reducing ductility at room temperature.

γ-TiAl alloys are outstanding materials, since suitable metallurgical processes can overcome their intrinsic brittleness. Therefore, the successful implementation of γ-TiAl alloys depends on achieving a suitable combination of room temperature ductility, mechanical strength, fatigue strength, creep resistance, fracture toughness and oxidation and corrosion resistances.

The mechanical properties obtained are strongly related to many factors, such as chemical composition, microstructure and processing technologies. Strict control of the processing conditions is essential in achieving the desired properties. Alloying can be one option for improving the ductility and fracture toughness of these alloys at high temperatures.

γ-TiAl alloys can be produced by conventional techniques such as ingot metallurgy, investment casting and powder metallurgy. The major requirement is the use of cost-effective processing techniques with proper control of microstructure.

The industrial implementation of these alloys also depends on the development of joining technologies. γ-TiAl joining is very difficult using fusion-welding processes due to their high reactivity and tendency to form brittle intermetallic phases. The development of cost-effective joining processes that allow the production of reliable joints is fundamental for expanding the use of γ-TiAl alloys to other components of the automotive and aerospace industries. In addition to similar joints, the joining of these alloys to steel and superalloys also has potential industrial applications.

Research and development on γ-TiAl alloys have progressed significantly in the recent decades. This chapter presents a review of the processing technologies and applications of these alloys. The main objective is the understanding of the relationship between the microstructure, chemical composition, processing techniques and mechanical properties of the γ-TiAl alloys.

1.2 Structure of γ-TiAl Alloys

The physical metallurgy of γ-TiAl alloys is very challenging. The microstructure and chemical composition are important factors that must be taken into account during the processing of these alloys, in order to obtain the desired properties.

The binary phase diagram Ti-Al contains several intermetallic compounds. In addition to the terminal solid solutions, α-Ti (with a hexagonal close-packed structure), β-Ti (with a body-centered cubic structure) and Al (with a faced-centered cubic structure), several intermetallic compounds are present: α_2-Ti_3Al, γ-TiAl, $TiAl_2$ and $TiAl_3$. Compared to pure titanium, the intermetallic compounds exhibit higher strength, due to their high ordering, and lower density depending on the aluminum content. Among these, only α_2-Ti_3Al and γ-TiAl have been widely studied, since $TiAl_2$ and $TiAl_3$ are extremely brittle. γ-TiAl with a tetragonal L10 structure (Figure 1.1) has

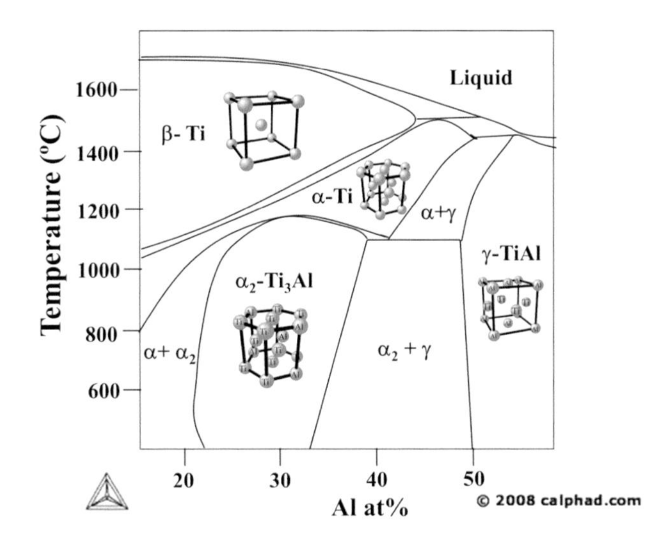

Figure 1.1 Ti-Al phase diagram obtained using CALPHAD and the representation of crystal structures of principal phases (adapted from CALPHAD website 2016).

been most closely studied, since it keeps its ordered structure up to higher temperatures, exhibits excellent oxidation resistance and has a lower density. The major drawback of this intermetallic compound is its low ductility at room temperature, for even with low levels of interstitial impurities it tends to fracture with an elongation of less than 1.0 per cent. Furthermore, the γ-TiAl alloys are susceptible to grain growth, which leads to a degradation of mechanical properties with increasing temperature. The presence of small amounts of the α_2-phase in the γ-TiAl matrix allows improving the alloys ductility (Appel and Oehring 2003, Lütjering and Williams 2007, Appel et al. 2011, Kothari et al. 2012, Clemens and Mayer 2013, Froes 2015). This improvement is related to the preferential dissolution of interstitial impurities in the α_2-phase, thus increasing the purity of the γ-phase, and to the refining of the microstructure. Therefore, the development of two-phase alloys has become attractive.

The solidification and solid-state phase transformations during cooling depends on alloy composition, as can be seen from the equilibrium phase diagram of Figure 1.1. Careful control of alloy composition and processing conditions are essential for adjusting the microstructure, thereby determining the final properties (Kothari et al. 2012, Clemens and Mayer 2013, Froes 2015).

Research and development activities have led to the introduction of two-phase alloys based on γ-TiAl with appropriate combinations of alloying elements to overcome the shortcomings in the mechanical properties of these alloys. The effect of alloying elements in the mechanical properties of two-phase alloys has been extensively investigated in order to improve the performance, particularly the creep resistance, room temperature ductility and oxidation resistance (Clemens and Kestler 2000, Kumpfert 2001, Gerling et al. 2004a,b, Wu 2006, Shu et al. 2014, Tian et al. 2014, Zhang et al. 2014). For instance, enhancement of room temperature ductility can be achieved by the addition of β-Ti stabilising elements, such as niobium, molybdenum, tantalum and vanadium.

The development of the so called second generation alloys was based on compositions close to Ti-(45-48)Al-(1-3)X-(2-5)Y-(<1)Z at.%, where X, Y and Z designate alloying elements: X = Cr, Mn or V; Y = Nb, Ta, W or Mo and Z = Si, B or C. The purpose of the adding of X elements is to improve the ductility. The adding of Y elements can increase the oxidation and creep resistances at high temperatures, while Z elements act as grain refiners (Clemens and Kestler 2000, Kumpfert 2001, Gerling et al. 2004a,b, Wu 2006, Stark et al. 2008, Wang et al. 2008, Cui and Liu 2009, Appel et al. 2014, Hu 2014, Shu et al. 2014, Tian et al. 2014).

Niobium and molybdenum are strong β-Ti stabilizing elements and considerably alter the phase boundaries of the Ti-Al phase diagram: the α-transus and β-Ti phase field are shifted to the aluminum rich side, the α_2-Ti_3Al phase is stabilised and the eutectoid temperature is raised. As a result of these alterations, the solidification path of the alloys changes: alloys with 45 per cent of aluminum solidify as β-Ti ($L \rightarrow \beta$), forming a structure of equiaxed grains, while for a typical Ti-50Al at.% alloy the solidification begins with the formation of α-Ti dendrites, with a preferential [0001] orientation parallel to the heat flow direction, and ends with a peritectic reaction ($L + \alpha \rightarrow \gamma$). Thus the beta solidification alloys present a refined structure, with no noticeable textures or segregation typical of the peritectic reactions (Zhang et al. 2014).

These investigations led to the development of several classes of high strength alloys rich in niobium and molybdenum (for instance, TNB, TNM and γ-Md alloys), also called third generation alloys. The niobium is responsible for improving the properties of γ-TiAl alloys: it lowers the stacking fault energy, increasing the deformation by twining and thus the room temperature ductility. It limits diffusion, improves creep resistance and increases oxidation resistance. These alloys were predominantly designed and optimised for hot forming, benefiting from properties associated with the high volume fraction of the β-Ti phase at high temperature. TNB alloys are Ti-45Al at.% with 5–10 at.% of molybdenum and small additions of carbon and boron used as grain refiners. High strength TNM alloys have equilibrated niobium and molybdenum concentration. γ-Md alloys are also niobium rich Ti-42Al at.% alloys; the high niobium and low aluminum content increase the β-Ti phase volume fraction and produce a β-Ti structure modulated at the nanoscale (Cui et al. 2007, Lin et al. 2007, Stark et al. 2008, Wang et al.

2008, Kothari et al. 2012, Clemens and Mayer 2013, Appel et al. 2014, Zhang et al. 2014).

The microstructure of γ-TiAl alloys can be classified as quasi-γ, duplex, nearly lamellar and fully lamellar. The alloys with a coarse fully lamellar microstructure have a relatively good fracture toughness and excellent creep resistance, but their ductility and tensile strength are low, especially at room temperature. On the other hand, alloys with fine-equiaxed grains, quasi-γ and duplex microstructures with small amounts of lamellar colonies have low fracture toughness and creep resistance, but moderated ductility and tensile strength at room and high temperatures. Hence, over recent decades, the development of alloys and processing routes that produce microstructures combining the best characteristics of duplex and lamellar microstructures has been one of the areas of research into γ-TiAl alloys (Appel and Oehring 2003, Lin et al. 2007, Lütjering and Williams 2007, Wang et al. 2008, Appel et al. 2011, Kothari et al. 2012, Clemens and Mayer 2013, Tian et al. 2014, Froes 2015).

Figure 1.2 shows the microstructures of two γ-TiAl alloys with different composition. The γ-TiAl alloy shown in Figure 1.2 (a) and (b) is a conventional alloy (Ti-51Al at.%) produced by arc melting. The microstructure of this alloy revealed the presence of lamellar two-phase dendrites ($\gamma + \alpha_2$) and interdendritic γ regions. The γ-TiAl alloy, whose microstructure can be seen in Figure 1.2 (c) and (d), belongs to a group called TNB v5. This alloy was produced by vacuum arc remelting and has a duplex microstructure with some γ grains. The TNB alloys can be considerably refined by hot working and annealing at low temperature, forming duplex or lamellar structures. The latter structures presenting some lamellas of the B19 phase that originated from the decomposition of the β-Ti phase or by additional ordering of the α_2-Ti_3Al phase. The γ-Md alloys structure is quite different, comprising slabs of the B19 phase intercalated with slabs of the β-Ti and α_2 phases. The B19 phase presents a nanometric domain structure (Clemens and Mayer 2013, Appel et al. 2014).

Thermo-mechanical processing and heat-treating have a strong influence on the microstructure, in particular on the γ/α_2 volume fraction. As for conventional materials, solid solution and precipitation hardening can be used to improve the mechanical properties. In the case of γ-TiAl alloys, it is important to achieve a compromise between the improvement in strength at elevated temperatures and room temperature ductility and toughness. Therefore, it is necessary to select the microstructure that optimises specific requirements in service (Cui et al. 2007, Lin et al. 2007, Wang et al. 2008, Cui and Liu 2009, Appel et al. 2014, Tian et al. 2014).

To sum up, to achieve the mechanical properties required for a given application of γ-TiAl alloys it is necessary to control production and thermo-mechanical processing conditions and maintain the best compromise

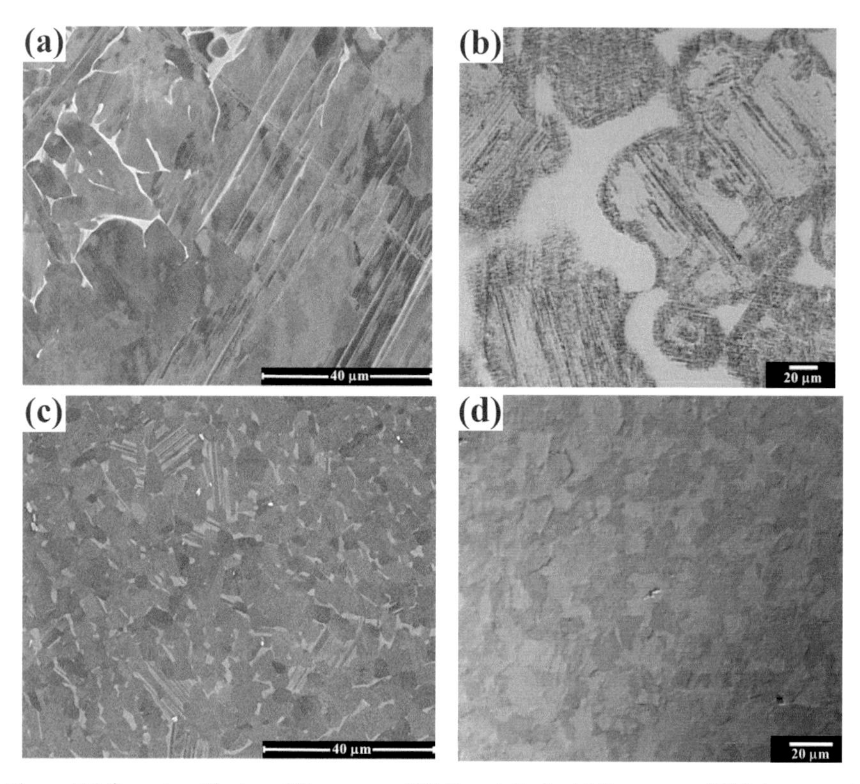

Figure 1.2 Scanning Electron Microscopy (SEM) and Optical Microscopy (OM) images of different γ-TiAl alloys with (a) and (b) Ti-51Al and (c) and (d) Ti-45Al-5Nb at.%.

between chemical composition and microstructure. However, the current limitations are due to the complex metallurgy of these alloys, resulting in a large scattering of mechanical properties as a consequence of chemical and structural inhomogeneity. Several investigations have been carried out in recent years to overcome these challenges of the γ-TiAl alloys and develop more attractive alloys with superior mechanical properties.

1.3 Properties of γ-TiAl Alloys

γ-TiAl alloys exhibit high specific strength, good corrosion and oxidation resistance, low creep at high temperatures and good fatigue resistance. These properties, associated with low density (3.9 to 4.1 g/cm^3), a high melting point (1,460°C), good structural stability and high-burn resistance make them attractive for high temperature industrial applications in which it is important to reduce weight. In these applications, the γ-TiAl alloys outweigh the disadvantages of conventional titanium alloys and compete directly with steels and nickel-based superalloys (Appel and Oehring 2003, Lütjering and Williams 2007, Appel et al. 2011, Kothari et al. 2012, Clemens and Mayer 2013, Froes 2015).

The comparison of physical and mechanical properties of the γ-TiAl alloys and conventional structural materials enables us to highlight the potential of these alloys for replacing conventional structural materials. In Table 1.1 some properties of γ-TiAl alloys, titanium alloys and nickel superalloys are listed. For temperatures between 600 and 800°C, the γ-TiAl alloys exhibit higher specific strength than the titanium alloys. Compared with nickel superalloys, the specific strength is very similar but the density is significantly lower. The specific Young's modulus of these alloys is 50 to 70 per cent higher than that of conventional titanium alloys and superalloys up to high temperatures (Appel and Oehring 2003, Lütjering and Williams 2007, Appel et al. 2011, Kothari et al. 2012, Clemens and Mayer 2013, Froes 2015).

In the last few years, modification of the chemical composition has been investigated in order to improve the mechanical properties of these alloys. As has already been mentioned, new generations of γ-TiAl alloys were developed. These new alloys, with appropriate thermo-mechanical processing and subsequent heat treatment, may have strength values

Table 1.1 Mechanical properties of titanium alloys, γ-TiAl alloys and nickel-based superalloys.

Property	Ti Alloys	γ-TiAl alloys	Ni-based superalloys
Density (g/cm^3)	4.5	3.7–3.9	7.9–9.2
Young's Modulus (GPa)	96–117	160–180	195–215
Yield Strength (MPa)	380–1,150	400–1,100	301–1,410
Tensile Strength (MPa)	480–1,200	450–1,150	517–1,750
% Ductility (RT)	2–20	1–4	3–5
% Ductility (°C)	12–50 (550°C)	10–60 (870°C)	20–80 (870°C)
Specific 1,000 hr Rupture Strength ($MPa/(g/cm^3)$) (°C)	100 (240°C) 60 (330°C)	100 (590°C) 60 (680°C)	100 (540°C) 60 (600°C)
Maximum Service Temperature (°C)	600	1,000	1,090
Oxidation (°C)	600	900	1,090
Temperature Significant Creep Deformation (°C)	315	650–700	650
Specific Modulus ($GPa/(g/cm^3)$)	25–19 (RT) 25–19 (600°C)	45–40 (RT) 30–25 (800°C)	35–20 (RT) 25–20 (800°C)
Specific Tensile Strength ($MPa/(g/cm^3)$)	135–200 (RT) 50–120 (600°C)	175–200 (RT) 150 (800°C)	125–150 (RT) 150–125 (800°C)

RT: Room temperature.
Data from Boyer et al. 1994, Wu 2006, Appel et al. 2008, 2011 and Jabbar et al. 2010.

superior to 1,000 MPa (Ciu et al. 2007, Lin et al. 2007, Wang et al. 2008, Appel et al. 2014).

For applications in which the main requirements are density and high service temperatures, these alloys can provide some advantages and compete with conventional structural materials. Despite recent advances, their low fracture toughness and fatigue resistance remain a problem for the design of some components. However, the advances made by the NASA Glenn Research Center, Plansee (Austria), the GKSS Research Center (Germany) and the University of Science and Technology Beijing (China) in recent years in the development of these alloys have been essential for expanding their industrial applications.

1.4 Processing of γ-TiAl Alloys

As already mentioned, processing of the γ-TiAl alloys will affect the final microstructure and thus the mechanical properties. During solidification and subsequent cooling to room temperature or during thermal or thermomechanical treatment, γ-TiAl alloys undergo several phase transformations. With the proper selection of production technology and optimisation of processing parameters, the required microstructure and properties can be achieved (Kestler and Clemens 2003, Couret et al. 2008, Cui and Liu 2009, Jabbar et al. 2010, 2011, Bolz et al. 2015).

The processing techniques on an industrial scale applied to the production of γ-TiAl components or semi-finished products have been recently reviewed (Kestler and Clemens 2003, Appel et al. 2011, Clemens and Mayer 2013). Conventional processing techniques, such as ingot metallurgy, investment casting and powder metallurgy are used for the production of the γ-TiAl alloys. However, the poor ductility of these alloys at room temperature makes it difficult to apply conventional cold working and machining processes, which makes near-net shape processes more competitive.

The ingots produced by casting processes require further processing, such as aging or hot working, to obtain the desired microstructure and chemical homogeneity of the alloys. In fact, the microstructure of these ingots consists of coarse lamellar grains of the α_2/γ phases due to the low cooling rate used in the industrial production of large ingots. A fine-grained microstructure may be obtained by centrifugal casting, where cooling rates are high, or when using β-Ti solidifying alloys. The as-cast ingots can be hot-worked, above the brittle-to-ductile transition temperature, to produce a fine-grained equiaxed microstructure consisting of near γ and/or lamellar phase morphology due to the dynamic recrystallization that occurs during forming operations (Couret et al. 2008, Cui and Liu 2009, Jabbar et al. 2010, 2011, Bolz et al. 2015). Through these additional processes the microstructure can be tailored in order to obtain the desired mechanical properties or to prepare the material for further thermomechanical treatments.

Bolz et al. (2015) applied two-step heat treatments to a forged β-Ti solidified alloy, an initial high-temperature treatment (near α-transus) and a subsequent annealing treatment. The microstructure consisted of lamellar (α_2/γ) colonies, equiaxed γ grains and β/β_0 grains (β_0 is ordered intermetallic phase with B2 structure that forms in alloys rich in β-Ti stabilising elements). The temperature and cooling rates of the high-temperature treatment exert a significant influence on the final microstructure and hence on the mechanical properties. The lower temperatures and slower cooling rates (furnace cooling) resulted in higher fractions of γ grains, while higher temperatures and more rapid cooling rates (air-cooling and oil-quenching) significantly increased the fraction of lamellar colonies.

Investment casting, including gravitational and centrifugal processes, has been suggested as the best way to produce near-net-shape γ-TiAl components at lower production costs (Kuang et al. 2000, 2001). This process involves little or no machining, thus avoiding a major limitation of these alloys. Several efforts have been made to select an appropriate investment casting process to produce high quality castings (Aguilar et al. 2011). However, the production of γ-TiAl parts by investment castings requires demanding and complex processes. Further process development is still required to overcome the problems associated with the high reactivity of the molten metal, the high melting temperature and poor fluidity, particularly in the selection of suitable mold face coat materials.

Powder Metallurgy (PM) is an alternative to casting processes since fine, homogeneous and non-textured microstructures are obtained without further processing (Thomas et al. 2005). However, the production of γ-TiAl powders is a very demanding and expensive process requiring the use of special techniques (Gerling et al. 2004a,b). Alternatively, elementary or mechanically milled powders have been used to produce γ-TiAl alloys by PM routes. Pre-alloyed γ-TiAl powders produce components with superior mechanical properties, since they minimise segregation and originate more homogeneous microstructures (Appel et al. 2011). However, there are studies claiming (Mei and Miyamoto 2001) that sintering of elementary powders is more effective and has better mechanical properties.

Several PM processing routes have been used to produce γ-TiAl alloys (Schloffer et al. 2012, Bewlay et al. 2013). Rapid sintering or consolidation techniques, among which Spark Plasma Sintering (SPS) stands out, have shown several advantages in the production of γ-TiAl alloys with the desired mechanical properties. Couret et al. (2008) studied the microstructure and mechanical properties of pre-alloyed γ-TiAl powders consolidated by SPS at temperatures between 1,100 and 1,250°C. The microstructure depended on the composition of the powders and the sintering temperatures. Samples

Table 1.2 Mechanical properties of γ-TiAl alloys produced by different processing technologies.

Chemical composition (at.%)	Processing technology			YS (MPa)	UTS (MPa)	E (GPa)
	Process	Temperature (°C)	Heat treatment			
Ti-47Al-2Cr-2Nb (Couret et al. 2008)	SPS	1,100	—	570	682	155
		1,150	—	563	654	166
		1,175	—	476	647	155
		1,190	—	472	569	155
		1,225	—	446	551	157
Ti-44Al-2Cr-2Nb-1B (Couret et al. 2008)	SPS	1,190	—	735	909	155
		1,225	—	712	881	159
		1,250	—	724	836	156
Ti-47Al-1Re-1W-0.2Si (Jabbar et al. 2010, 2011)	SPS	1,275	—	580	620	193
		1,225	—	700	720	175
Ti-45Al-2Nb-2Cr **Ti-45Al-6Nb-1Mo-0.1B** (Bolz et al. 2015)	Casting + forging + heat treatment	Remelting a VAR Ingot + HIP (1,210°C/200 MPa/4 hr) + Forging (1,270°C)	1,300°C/1 hr/air cooling + 800°C/6 hr	—	900	175
			1,270°C/1 hr/air cooling + 800°C/6 hr		950	175
			1,250°C/1 hr/air cooling + 800°C/6 hr		900	175
			1,270°C/1 hr/oil quenching + 800°C/6 hr		1,000	194
			1,270°C/1 hr/furnace cooling + 800°C/6 hr		810	175

VAR: Vacuum Arc Melting; HIP: Hot Isostatic Pressing; YS: Yield Strength; UTS: Ultimate Tensile Strength; E: Young's Modulus.

sintered at lower temperatures (1,100 and 1,150°C) exhibited a microstructure composed of $\gamma+\alpha_2$ grains. Increasing the sintering temperature led to a change to a lamellar microstructure. Higher mechanical properties at room temperature were observed in samples with duplex microstructure, while the lamellar microstructure had better creep resistance at 700°C. Jabbar et al. (2010, 2011) also demonstrated the feasibility of using the SPS technique to produce TNB and G4 (Ti-47Al-1Re-1W-0.2Si at.%) alloys. However, the use of these powders made microstructural control difficult and further thermal treatments were needed to improve mechanical properties.

Selective Electron Beam Melting (SEBM) and Selective Laser Melting (SLM) are two routes used to produce γ-TiAl by additive manufacturing (Cormier et al. 2007, Loeber et al. 2011, Gussone et al. 2015, Li, D. et al. 2016, Li, W. et al. 2016). Although the advantages of additive manufacturing are recognised, particularly for the production of small series of complex structures, these techniques require further development before the production of γ-TiAl components of the desired quality can be achieved.

The mechanical properties of several γ-TiAl alloys produced by different processing technologies are listed in Table 1.2. These results show the dependence of the properties on manufacturing process and composition. The alloys produced by casting and subjected to forging process and subsequent heat treatments have the highest mechanical properties.

To sum up, the manufacturing technologies significantly influence the microstructure and mechanical properties of the γ-TiAl alloys. Many studies have been focused on developing the desired microstructure in order to obtain the most appropriate mechanical properties. It has been demonstrated that a strict control of production conditions and the development of new production technologies is crucial in obtaining new γ-TiAl alloys with more attractive mechanical properties.

1.5 Concluding Remarks

γ-TiAl alloys are considered materials for replacing traditional metallic alloys, especially when weight reduction of the components is a critical factor for structural application. These alloys exhibit a range of attractive properties for the aerospace and automotive industries such as low density, high temperature strength, good resistance against oxidation and corrosion, and high burn resistance. The successful development of manufacturing and joining processes is very important for increasing the applications of these alloys. Several studies have been conducted in order to understand and relate the chemical composition and microstructure of these alloys with the manufacturing processes and mechanical properties. γ-TiAl alloys can be produced through conventional processes, but more recent or adapted processes, combined with heat treatments, are those that allow better control of the final microstructure.

Joining technologies are also a key to the implementation of these alloys. Some components produced with these alloys may need to be bonded to steel and nickel-based superalloys. The joining of the γ-TiAl alloys to themselves or to other materials is very difficult by conventional processes and approaches to the joining technologies need to be developed. This subject will be enlarged in the following chapters with a special focus on solid-state diffusion bonding.

Keywords: Aerospace; automotive; casting; joining process; γ-TiAl; manufacturing; mechanical properties; microstructure; powder metallurgy; spark plasma sintering; selective electron beam melting; selective laser melting; steels; superalloys; titanium alloys.

1.6 References

Aguilar, J., A. Schievenbusch and O. Kättlitz. 2011. Investment casting technology for production of TiAl low pressure turbine blades—Process engineering and parameter analysis. Intermetallics 19: 757–761.

Appel, F. and M. Oehring. 2003. γ-Titanium aluminide alloys: alloy design and properties. pp. 89–146. *In:* C. Leyens and M. Peters [eds.]. Titanium and Titanium Alloys: Fundamentals and Applications. Wiley-VCH Verlag GmbH & Co. KGaA, Weinheim, Germany.

Appel, F., M. Oehring and J.D.H. Paul. 2008. A novel *in situ* composite structure in TiAl alloys. Mater. Sci. Eng. A-Struct. Mater. Prop. Microstruct. Process. 493: 232–236.

Appel, F., J.D.H. Paul and M. Oehring. 2011. Gamma Titanium Aluminide Alloys: Science and Technology. Wiley-VCH Verlag GmbH & Co. KGaA, Weinheim, Germany.

Appel, F., M. Oehring and J.D.H. Paul. 2014. Physical metallurgy and performance of the TNB and γ-Md alloys. pp. 9–20. *In:* Y.-W. Kim, W. Smarsly, J. Lin, D. Dimiduk and F. Appel [eds.]. Gamma Titanium Aluminide Alloys 2014. John Wiley & Sons, Inc, Hoboken, New Jersey, USA.

Bewlay, B.P., M. Weimer, T. Kelly, A. Suzuki and P.R. Subramanian. 2013. The science, technology and implementation of TiAl alloys in commercial aircraft engines. MRS Symp. Proc. 1516: 49–58.

Bolz, S., M. Oehring, J. Lindemann, F. Pyczak, J. Paul, A. Stark et al. 2015. Microstructure and mechanical properties of a forged β-solidifying γ-TiAl alloy in different heat treatment conditions. Intermetallics 58: 71–83.

Boyer, R., G. Welsch and E.W. Collings [eds.]. 1994. Materials Properties Handbook - Titanium Alloys. ASM International, Materials Park, OH, USA.

CALPHAD website. 2016. Computational Thermodynamics: Calculation of phase diagrams using the CALPHAD method. http://www.calphad.com/titanium-aluminum.html.

Clemens, H. and H. Kestler. 2000. Processing and applications of intermetallic γ-TiAl-based alloys. Adv. Eng. Mater. 2: 551–570.

Clemens, H. and S. Mayer. 2013. Design, processing, microstructure, properties, and applications of advanced intermetallic TiAl alloys. Adv. Eng. Mater. 15: 191–215.

Cormier, D., O. Harrysson, T. Mahale and H. West. 2007. Freeform fabrication of titanium aluminide via electron beam melting using prealloyed and blended powders. Res. Lett. Mater. Sci. 2007: 1–4.

Couret, A., G. Molénat, J. Galy and M. Thomas. 2008. Microstructures and mechanical properties of TiAl alloys consolidated by spark plasma sintering. Intermetallics 16: 1134–1141.

Cui, W.F., C.M. Liu, V. Bauer and H.-J. Christ. 2007. Thermomechanical fatigue behaviours of a third generation γ-TiAl based alloy. Intermetallics 15: 675–678.

Cui, W.F. and C.M. Liu. 2009. Fracture characteristics of γ-TiAl alloy with high Nb content under cyclic loading. J. Alloy. Compd. 477: 596–601.
Dimiduk, D.M. 1999. Gamma titanium aluminide alloys—an assessment within the competition of aerospace structural materials. Mater. Sci. Eng. A-Struct. Mater. Prop. Microstruct. Process. 263: 281–288.
Froes, F.H. [ed.]. 2015. Titanium: Physical Metallurgy, Processing and Application. ASM International, Materials Park, OH, USA.
Gerling, R., A. Bartels, H. Clemens, H. Kestler and F.-P. Schimansky. 2004a. Structural characterization and tensile properties of a high niobium containing gamma TiAl sheet obtained by powder metallurgical processing. Intermetallics 12: 275–280.
Gerling, R., H. Clemens and F.-P. Schimansky. 2004b. Powder metallurgical processing of intermetallic gamma titanium aluminides. Adv. Eng. Mater. 6: 23–38.
Gussone, J., Y.-C. Hagedorn, H. Gherssekhloo, G. Kasperovich, T. Merzouk and J. Hausmann. 2015. Microstructure of γ-titanium aluminide processed by selective laser melting at elevated temperatures. Intermetallics 66: 133–140.
Hu, D. 2014. A quarter century journey of boron as a grain refiner in TiAl alloys. pp. 21–30. *In:* Y.-W. Kim, W. Smarsly, J. Lin, D. Dimiduk and F. Appel [eds.]. Gamma Titanium Aluminide Alloys 2014. John Wiley & Sons, Inc, Hoboken, New Jersey, USA.
Jabbar, H., J.-P. Monchoux, F. Houdellier, M. Dollé, F.-P. Schimansky, F. Pyczak et al. 2010. Microstructure and mechanical properties of high niobium containing TiAl alloys elaborated by spark plasma sintering. Intermetallics 18: 2312–2321.
Jabbar, H., J.-P. Monchoux, M. Thomas and A. Couret. 2011. Microstructures and deformation mechanisms of a G4 TiAl alloy produced by spark plasma sintering. Acta Mater. 59: 7574–7585.
Kestler, H. and H. Clemens. 2003. Production, processing and application of γ(TiAl)-based alloys. pp. 351–388. *In:* C. Leyens and M. Peters [eds.]. Titanium and Titanium Alloys: Fundamentals and Applications. Wiley-VCH Verlag GmbH & Co. KGaA, Weinheim, Germany.
Knippscheer, S. and G. Frommeyer. 1999. Intermetallic TiAl (Cr, Mo, Si) alloys for lightweight engine parts: structure, properties and applications. Adv. Eng. Mater. 1: 187–191.
Kothari, K., R. Radhakrishnan and N.M. Wereley. 2012. Advances in gamma titanium aluminides and their manufacturing techniques. Prog. Aerosp. Sci. 55: 1–16.
Kuang, J.P., R.A. Harding and J. Campbell. 2000. Examination of defects in gamma titanium aluminide investment castings. Int. J. Cast. Metals Res. 13: 125–134.
Kuang, J.P., R.A. Harding and J. Campbell. 2001. A study of refractories as crucible and mould materials for melting and casting γ-TiAl alloys. Int. J. Cast. Metals Res. 13: 277–292.
Kumpfert, J. 2001. Intermetallic alloys based on orthorhombic titanium aluminide. Adv. Eng. Mater. 3: 851–864.
Li, D., S. Hu, J. Shen, H. Zhang and X. Bu. 2016. Microstructure and mechanical properties of laser-welded joints of Ti-22Al-25Nb/TA15 dissimilar titanium alloys. J. Mater. Eng. Perform. 25: 1880–1888.
Li, W., J. Liu, Y. Zhou, S. Wen, Q. Wei, C. Yan et al. 2016. Effect of substrate preheating on the texture, phase and nanohardness of a Ti-45Al-2Cr-5NB alloy processed by selective laser melting. Scr. Mater. 118: 13–18.
Lin, J.P., X.J. Xu, Y.L. Wang, S.F. He, Y. Zhang, X.P. Song et al. 2007. High temperature deformation behaviors of a high Nb containing TiAl alloy. Intermetallics 15: 668–674.
Loeber, L., S. Biamino, U. Ackelid, S. Sabbadini, P. Epicoco, P. Fino et al. 2011. Comparison of selective laser and electron beam melted titanium aluminides. Solid Freeform Fabr. Symp. Proc. USA 547–556.
Lütjering, G. and J.C. Williams. 2007. Titanium. 2nd edition Springer-Verlag Berlin, Heidelberg, Germany.
Mei, B. and Y. Miyamoto. 2001. Preparation of Ti-Al intermetallic compounds by spark plasma sintering. Metall. Mater. Trans. A-Phys. Metall. Mater. Sci. 32A: 843–847.

Peters, M., J. Hemptenmacher, J. Kumpfert and C. Leyens. 2003. Structure and properties of titanium and titanium alloys. pp. 1–35. *In:* C. Leyens and M. Peters [eds.]. Titanium and Titanium Alloys: Fundamentals and Applications. Wiley-VCH Verlag GmbH & Co. KGaA, Weinheim, Germany.
Schloffer, M., F. Iqbal, H. Gabrisch, E. Schwaighofer, F.-P. Schimansky, S. Mayer et al. 2012. Microstructure development and hardness of a powder metallurgical multi-phase γ-TiAl based alloy. Intermetallics 22: 231–240.
Shu, S., F. Qiu, C. Tong, X. Shan and Q. Jiang. 2014. Effects of Fe, Co and Ni elements on the ductility of TiAl alloys. J. Alloy. Compd. 617: 302–305.
Stark, A., A. Bartels, H. Clemens and F.-P. Schimansky. 2008. On the formation of ordered ω-phase in high Nb containing γ-TiAl based alloys. Adv. Eng. Mater. 10: 929–934.
Thomas, M., J.L. Raviart and F. Popoff. 2005. Cast and PM processing developments in gamma aluminides. Intermetallics 13: 944–951.
Tian, S.G., Q. Wang, H.C. Yu, H.F. Sun and Q.Y. Li. 2014. Influence of heat treatment on microstructure and creep properties of TiAl-Nb alloy. Mater. Res. Innov. 18: S4336–S4340.
Wang, Y.H., J.P. Lin, Y.H. He, X. Lu, Y.L. Wang and G.L. Chen. 2008. Microstructure and mechanical properties of high Nb containing TiAl alloys by reactive hot pressing. J. Alloy. Compd. 461: 367–372.
Wu, X. 2006. Review of alloy and process development of TiAl alloys. Intermetallics 14: 1114–1122.
Zhang, L., J. Zheng, Y. Hou, X. Ma, X. Xu and J. Lin. 2014. Composition optimization of β-γ TiAl alloys containing high niobium. pp. 31–38. *In:* Y.-W. Kim, W. Smarsly, J. Lin, D. Dimiduk and F. Appel [eds.]. Gamma Titanium Aluminide Alloys 2014. John Wiley & Sons, Inc, Hoboken, New Jersey, USA.

CHAPTER 2

Joining of γ-TiAl Alloys

2.1 Introduction

The industrial applications of γ-TiAl alloys on a large scale require the development of joining techniques producing joints of good quality and high mechanical properties, suitable for the performance of components in service (Kestler and Clemens 2003, Messler Jr. 2004a,b, Lütjering and Williams 2007, Appel et al. 2011). Joining techniques can provide secure and dependable means of combining materials, thus assuring mechanical coupling or support, electric connection or insulation, environmental protection, etc. In recent years, various studies have been conducted to investigate joining techniques of these alloys with the aim of establishing methods which are industrially feasible. These alloys have limitations on weldability, requiring strict procedures to achieve a good quality of the joints. These limitations are mostly due to their low ductility, which is associated with thermal residual stresses and the formation of brittle intermetallic phases at the joint that can lead to cracking.

γ-TiAl alloys can be joined to themselves and to other materials through various processes such as fusion welding (Mallory et al. 1994, Hirose et al. 1995, Chaturvedi et al. 1997, Xu et al. 1999, 2001, Arenas and Acoff 2002, 2003, Ranatowski 2008, Cao et al. 2014, Liu et al. 2014), friction welding (Miyashita and Hino 1994, Lee et al. 2004a,b, Ventzke et al. 2010), brazing (Humpston and Jacobson 1993, Lee et al. 1998, Lee and Wu 1999, Guedes et al. 2002, 2003, 2004, 2006, Shiue et al. 2004, Wallis et al. 2004, He et al. 2005, 2009, Song et al. 2012a,b, Mirski and Rozanski 2013), transient liquid phase bonding (Duan et al. 2004, Lin et al. 2013) and solid-state diffusion bonding (that will be described in detail in Chapter 3). Solid-state diffusion bonding and vacuum brazing have been most investigated for the possibility of producing reliable joints.

The main objective of this chapter is the presentation of the fundamental concepts and the advantages and disadvantages of each process in the joining of these alloys.

2.2 Fusion Welding

Fusion welding processes most reported as capable of joining γ-TiAl alloys are gas tungsten arc welding, laser welding and electron beam welding (Mallory et al. 1994, Hirose et al. 1995, Chaturvedi et al. 1997, 2001, Xu et al. 1999, Arenas and Acoff 2002, 2003, Ranatowski 2008, Cao et al. 2014, Liu et al. 2014). The fusion welding of the γ-TiAl alloys is quite difficult, due to the high reactivity of these alloys, the risk of solid-state cracking and the lack of suitable filler metals (Threadgill 1995). Despite all these difficulties, some satisfactory results were obtained with fusion welding methods (Xu et al. 1999, Chaturvedi et al. 2001, Arenas and Acoff 2003), but the properties of the welds were usually not ideal. Lower values of tensile strength, ductility and fatigue resistance have often been reported, and these rarely reach the values of base materials (Mallory et al. 1994, Ranatowski 2008, Cao et al. 2014). The melting, solidification and rapid cooling rates result in complex phase transformations that lead to the formation of microstructures containing high amounts of α_2-Ti_3Al, with a high concentration of residual thermal stresses. As already mentioned in Chapter 1, the mechanical properties are strongly related to the microstructure, so that it is crucial to understand and optimise the processing technology.

As gas tungsten arc welding is used only for components that do not require high integrity and quality, it is not used in components for the aerospace industry. The process can be used to weld γ-TiAl alloys up to 20 mm thick. However, components with thicknesses greater than 2 mm require two or more welding passes. In general, a good bond occurs when the hardness of the weld seam is slightly higher than the base material. Post-Bond Heat Treatment (PBHT) for stress relief is recommended when the samples are thick or involve complex geometry. As γ-TiAl alloys are very reactive, a protective gas shield must be provided at temperatures above 300ºC (Kestler and Clemens 2003, Messler Jr. 2004a,b, Lütjering and Williams 2007, Appel et al. 2011).

Arenas and Acoff (2003) studied the weldability of γ-TiAl alloys (Ti-48Al-2Nb-2Cr, at.%) using an autogenous gas tungsten arc welding process. The welding currents used ranged between 50 and 150 A. The base materials were not preheated, although a stress relief treatment at 615ºC for 120 min was performed before welding. Solidification cracking and solid-state cracking were observed. The base material had a structure composed of a lamellar (γ/α_2) structure with a few γ grains. This microstructure changed to columnar and equiaxed dendritic structures at the welding interface, where supersaturated α_2-Ti_3Al phase were also observed. The susceptibility to cracking decreased as the welding current was increased, which is correlated with the decrease in the α_2-Ti_3Al phase volume fraction. A sound joint was obtained with higher welding currents.

However, the mechanical properties were substantially lower than those of the base material, with a decrease of 50 per cent in the tensile strength.

The same authors (Arenas and Acoff 2002) reported that the mechanical properties of the γ-TiAl alloys joined by gas tungsten arc welding could be improved by PBHT. This treatment promoted a reduction in the amount of the α_2-Ti_3Al phase, which was transformed into the γ-TiAl phase, decreasing the hardness of the fusion zone to values close to those of the base material. However, to achieve this reduction in hardness, treatments at high temperatures (1,000–1,200°C) for long dwell times (from 600 to 2,880 min) were required.

Electron beam welding is the process most common for fusion welding of titanium aluminides. In this process the heat is generated by a beam of high-velocity electrons applied to the components to be joined. Contrary to gas tungsten arc welding, this process can be applied to joining γ-TiAl components for aerospace applications. Furthermore, with this process it is possible to obtain an excellent reproducibility of the quality of the joints. However, it is more expensive, since it is performed in high vacuum. The need to use a vacuum chamber is a limitation of the process since it restricts the size of the components to be welded (Kestler and Clemens 2003, Messler Jr. 2004a,b, Lütjering and Williams 2007, Appel et al. 2011). The solid-state cracking is the main problem when applying this process to join γ-TiAl alloys.

Chaturvedi et al. (1997, 2001) successfully used a conventional electron beam for welding small pieces of Ti-45Al-2Nb-2Mn + 0.8% vol. TiB_2 alloy. The authors investigated the influence of the cooling rate, welding speed, the amount of heat and the preheating temperature in the microstructural changes. To obtain a good quality joint by this process, it was necessary to use well-selected welding parameters, namely a cooling rate lower than 250 K/s. For high cooling rates the suppression of the α to γ phase transformation occurred; the welding region remained rich in the very brittle α_2-Ti_3Al phase, thus promoting solid-state cracking.

Xu et al. (1999) also investigated the possibility of joining γ-TiAl alloys by means of the electron beam welding process. Two alloys with different compositions were tested: Ti-45Al-2Nb-2Mn and Ti-48Al-2Nb-2Mn at.%. It was observed that these alloys were susceptible to solidification cracking due to high thermally induced stresses and the intrinsic brittleness of the microstructures. However, the Ti-48Al-2Nb-2Mn alloy was less susceptible to cracking as the α(Ti) phase decomposes more readily.

Chen et al. (2011) modified the electron beam welding, prolonging the high temperature stage by preheating the base material and using a ceramic as a thermal insulation underlay. With this modification of the welding parameters the authors obtained a microstructure composed of γ-TiAl grains and γ/α_2 colonies instead of the α_2-Ti_3Al structure obtained by the direct welding method. The resulting microstructure of this modified method reduces cold cracking and increases the tensile strength of the joints to 411 MPa (an increase of 20 per cent compared with the direct method).

Laser welding is another fusion welding process that can be used to join γ-TiAl alloys. Like electron beam welding, it is essentially characterised by being a high power density process. This technique offers some advantages in comparison to electron beam welding, such as higher speed, good automated manufacturing, high reproducibility and low risk of distortion of the joints. The limitations of this process are the difficulty of joining components with complex geometry and the possibility of the laser beam being deflected by lenses, mirrors or fiber optics. The Heat-Affected Zone (HAZ) is usually small and the heating and cooling rates are very high. The primary lasers commonly used are CO_2 (gas laser) and Nd:YAG (solid-state laser) (Kestler and Clemens 2003, Messler Jr. 2004a,b, Lütjering and Williams 2007, Appel et al. 2011, Liu et al. 2012).

Hirose et al. (1995) studied the applicability of the laser welding process to bonding Ti-46Al-2Mo at.% alloy using a CO_2 laser with a power of 2.5 kW. The authors investigated the effects of different cooling rates, which conditioned the microstructure obtained. Very high cooling rates led to the formation of a microstructure in the fusion zone composed of the hard, brittle α_2-Ti_3Al phase, while for lower cooling rate the structure was composed of massive α_2-Ti_3Al, massive γ-TiAl and lamellar γ-TiAl grains. The hardness value of the welding zone increased from 370 HV4.9 to 450 HV4.9 with an increase in the cooling rates—the weld zone is harder than the base material (315 HV4.9). This increase was related to the presence of the α_2-Ti_3Al phase at the fusion zone. The increase in the cooling rate also increased the crack sensitivity of the laser welds. Cracking was prevented by selecting optimum welding conditions, which were a cooling rate of less than 30 K/s between 800 and 600ºC and preheating temperatures above 300ºC. In tensile tests of samples welded under these conditions, the samples fractured in the base material, showing the laser welding quality.

Liu et al. (2012, 2014) laser welded γ-TiAl alloys (Ti-48Al-1Cr-1.5Nb-1Mn-0.2Si-0.5B, at.%) using a three steps approach: preheating at 750ºC, welding and PBHT. The authors used two Nd:YAG lasers and, in one case, a secondary defocused laser beam to carry out an *in situ* PBHT immediately after welding. This *in situ* treatment significantly reduced the residual stresses, but was not sufficient to achieve a good weld microstructure (Liu et al. 2012). A conventional PBHT at 1,260ºC for 120 min was needed to reduce the fraction of the α_2-Ti_3Al phase in the welding zone, thus improving the mechanical properties of the weld. Similar results can be obtained by replacing the two-steps PBHT with a conventional treatment performed at 1,350ºC for 30 min (Liu et al. 2014). Figure 2.1 shows the microstructures of the different zones of the

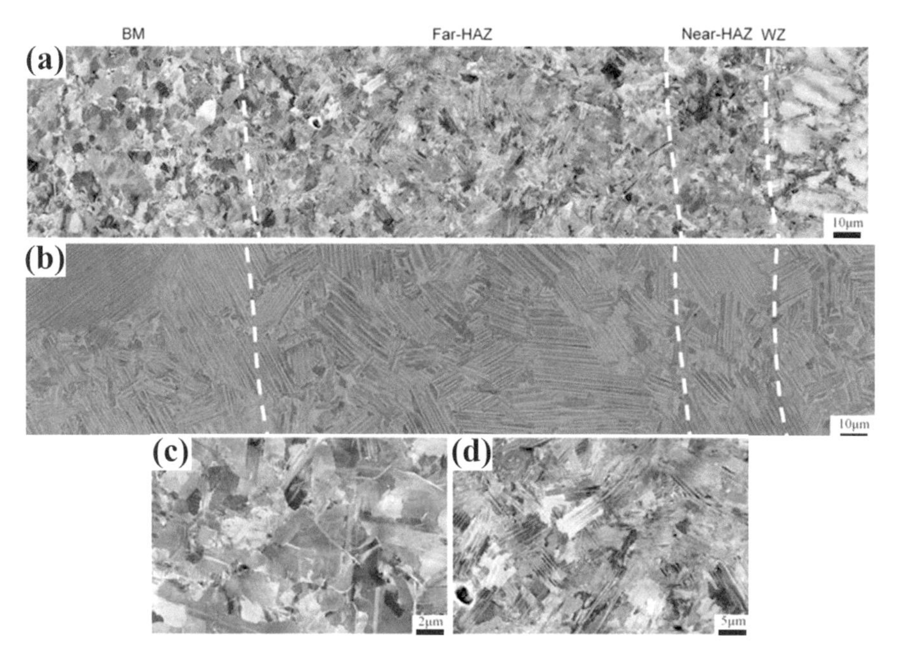

Figure 2.1 Microstructures of WZ, HAZ and Base Material (BM) in (a) as-welded and (b) post-bond heat treated specimens, and (c) and (d) near HAZ and far-HAZ of as-welded specimen. (Reprinted from Metall. Mater. Trans. A-Phys. Metall. Mater. Sci. 45A: 16–28. Liu, J., V. Ventzke, P. Staron, N. Schell, N. Kashaev and N. Huber. Effect of post-weld heat treatment on microstructure and mechanical properties of laser beam welded TiAl-based alloys. Copyright 2013, with permission from Springer.)

as-welded and PBHT samples. The weld zone (WZ) of the as-welded specimen was characterized by coarse lamellar dendrites. The PBHT relieved the residual stresses and promoted the formation of an almost fully lamellar microstructure (Figure 2.1 (b)). This microstructural change improved the tensile strength from 355 MPa to 524 MPa for the as-welded and PBHT samples, respectively.

Table 2.1 displays the mechanical properties obtained for some joints produced by fusion welding processes. A dramatic decrease in the mechanical properties is observed for welded samples with gas tungsten arc welding. This can be explained by the microstructural changes that occur during the process, promoting the formation of the brittle α_2-Ti_3Al phase in WZ. With the laser welding process the samples exhibited tensile strengths closer to the base material. This table also shows that the mechanical properties can be improved by a PBHT that reduces the fraction of the α_2 phase and promotes a more homogeneous microstructure.

Table 2.1 Tensile strength values of γ-TiAl joints produced by fusion welding processes.

γ-TiAl alloys composition (at.%)	**Fusion welding process**	**Tensile strength (MPa)**	
Ti-48Al-2Nb-2Cr (Arenas and Acoff 2002, 2003)	Gas tungsten arc welding	Base Material	400
		As-welded	240
Ti-46Al-2Mo (Hirose et al. 1995)	Laser welding	As-welded	450
Ti-48Al-1Cr-1.5Nb-1Mn-0.2Si-0.5B (Liu et al. 2014)	Laser welding	As-welded	355
		PBHT samples	524
Ti-43Al-9V-0.3Y (Chen et al. 2011)	Electron beam welding	As-welded	411

PBHT: Post-Bond Heat Treatment.

In summary, conventional or fusion welding processes, such as gas arc welding and laser welding, can be used for joining γ-TiAl alloys. A very tight control over all welding parameters is crucial in producing joints free of defects. Laser welding is the most promising fusion process for joining γ-TiAl alloys, since it allows the production of joints with better mechanical properties. However, the as-welded mechanical properties are low and must be enhanced by PBHT. The PBHT are not very attractive for industrial applications, since the high temperatures and long dwell time needed make them very expensive and time consuming. For these reasons, bonding processes that do not involve the fusion of the base material are more attractive for producing good-quality joints of γ-TiAl alloys.

2.3 Friction Welding

Friction welding is a process that uses the energy produced by the friction resulting from the rotation of the two pieces under pressure (Figure 2.2 (a)). The friction that occurs in this process leads to local heating between the two components to be joined, thus allowing the solid-state bonding. The high plastic deformation resulting from friction between components associated with the local increase in temperature causes an intense deformation and dynamic recrystallization process. Friction welding is classified as a solid-state welding process whereby metallic bonding is produced at temperatures lower than the melting point of the base materials. Friction time, friction pressure and rotation speed are the most important parameters in the friction welding process (Kestler and Clemens 2003, Messler Jr. 2004a, b, Lütjering and Williams 2007, Appel et al. 2011, Cao et al. 2014).

The process can also be applied by pressing a rotating tool against the two facing surfaces to be joined. In this case the process is called Friction Stir Welding (FSW)—see the schematic illustration of the processes in Figure 2.2 (b). Recently, friction welding has attracted special attention

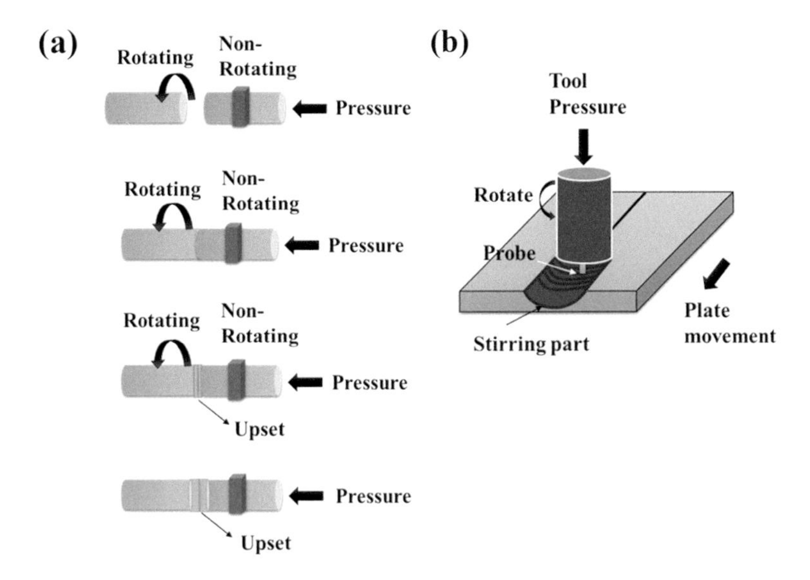

Figure 2.2 Schematic illustration of the: (a) friction welding and (b) friction stir welding processes.

regarding the production of rotors (disk blades) for jet engine compressors. Its low cost and the possibility of its being applied to the joining of large components are the main advantages of this process (Kestler and Clemens 2003, Messler Jr. 2004a,b, Lütjering and Williams 2007, Appel et al. 2011).

Three zones characterise the interface of the components joined by the friction welding process: the central zone, the Thermo-Mechanically Affected Zone (TMAZ) and HAZ. Processing conditions at the central zone (temperature and deformation rate) are sufficient to promote dynamic recrystallization. The material in the TMAZ also undergoes intense plastic deformation. However, it is insufficient to promote dynamic recrystallization. The HAZ is characterised by small microstructural changes. Nevertheless, in this region there can occur changes associated with local heat treatment (aging, tempering or annealing) (Kestler and Clemens 2003, Messler Jr. 2004a,b, Lütjering and Williams 2007, Appel et al. 2011).

The joining of γ-TiAl to itself by friction stir welding has been the subject of very few studies. The work of Miyashita and Hino (1994) is one of these few investigations. Samples of γ-TiAl with a near equiatomic composition and diameters of 6 and 12 mm were welded. Tensile specimens of the samples with 12 mm in diameter fractured through the base material with a tensile strength higher than 530 MPa. For the samples of 6 mm in diameter, the tensile specimens often fractured by the welding zone and only under well-controlled conditions did the fracture occur in the base material. Some micro-cracks were observed in the TMAZ that it was necessary to eliminate to improve the mechanical properties of the joint.

The joining of γ-TiAl to other materials by friction welding has been more widely reported. Since this welding process does not involve the melting of the base materials, it can be used for bonding dissimilar materials with quite different mechanical and physical properties.

Lee et al. (2004a) joined γ-TiAl (Ti-47Al at.%) and Fe-0.43C-1.09Cr-0.67Mn-0.31Si-0.2Mo wt.% steel rods, with 20 mm and 24 mm in diameter respectively, by friction welding. The friction welding parameters used were a rotating speed of 2,000 rpm, a friction time change from 30 to 50 s and a friction pressure between 130 and 170 MPa. The formation of brittle intermetallics at the interface led to the formation of cracks there. The tensile strength obtained was very low (120 MPa) due to these cracks. The cracking was explained by the internal stresses induced by the martensitic transformation of the steel due to the high cooling rate after joining. In order to improve the joint quality, the authors used pure copper as an intermediate layer (Lee et al. 2004b). In this case, joining was performed in two friction welding steps: first, the copper layer was joined to the steel and then this body to the γ-TiAl alloy. Copper acted as a stress relief layer, minimising the tendency to cracking and thus increasing the quality of the joints. In addition, the high thermal conductivity of copper contributed to a decrease in the HAZ in the side of the steel. This improvement in the quality of the bond interface was reflected in the mechanical response of the joints; the tensile strength increased to 375 MPa, and under optimised friction welding conditions, fracture occurred through the γ-TiAl base material.

The dissimilar friction welding of γ-TiAl to steel was also studied by Dong et al. (2005). γ-TiAl (Ti–43Al–2Cr–Zr–Fe at.%) and Fe-0.43C-1.09Cr-0.67Mn-0.31Si-0.2Mo wt.% steel rods were joined at a rotating speed of 1,500 rpm and a friction pressure of 186 MPa. The as-welded joint was very brittle and fractured during machining. PBHT at 580°C for 120 min reduced the internal stresses by tempering the martensite. This microstructural modification significantly improved the quality of the joint and a tensile strength of 405 MPa was measured; the fracture occurred through the γ-TiAl alloy. Intermetallic phases ($TiFe_2$ and γ-TiAl) and carbides (TiC) were observed at the interface.

Ventzke et al. (2010) investigated the friction welding of γ-TiAl (Ti-47Al-3.5-(Mn+Cr+Nb)-0.8(B+Si) at.%) to Ti6Al4V alloy. Specimens of 25 mm in diameter were used. The experiments were performed at a friction pressure of 310 MPa and an average process time of 4.7 s. The joints produced showed a large number of voids aligned along the interfaces, a reaction zone formed by the brittle α_2-Ti_3Al phase in the Ti6Al4V side, and cracks in the γ-TiAl side. Sun et al. (2012) used synchrotron X-ray computed tomography to inspect friction-welded joints in γ-TiAl turbocharger rotors. This component consisted of a cast γ-TiAl turbine wheel joined to a Ti6Al4V shaft. Three friction-welded rotors were tested and all revealed cracks at the welding interface and inside the γ-TiAl alloy.

Although this process does not involve the melting of the base material, the microstructural changes occurring due to the high plastic deformation and high cooling rates are significant and affect the mechanical behaviour of the welded specimens. These results strongly depend on the process conditions and the optimisation of all parameters for each type of joint is required. In some cases, it may be necessary to carry out PBHT or use ductile interlayers.

2.4 Brazing

Brazing is a joining process that does not involve the melting of the base materials. The process uses brazing filler that melts and wets the base material, filling the joint by means of capillary forces. The liquid filler reacts with the solid base material, dissolving it and forming a reaction layer at the joint. This reaction continues until solidification is achieved and must be controlled to avoid excessive dissolution of the base materials that can cause the quality of the joint to deteriorate. The advantages of this process are the low joining temperature and short dwell time, which reduce the tendency for solid-state cracking and enable the possibility of joining dissimilar materials and materials with complex geometries. The success of this process depends on several factors, including the characteristics of the base material and the filler material, the wettability of the base material by the filler, careful preparation of the surfaces, joints design, heat source and the atmosphere used (Humpston and Jacobson 1993, Lee et al. 1998, Cao et al. 2014).

There are some variations in the brazing process. For example, active metal brazing, in which a metal (usually titanium) is added to the filler alloy to promote reaction and wetting with the substrate. The use of active metal brazing is quite redundant in joining γ-TiAl, which is wet by most of the fillers, since the active element is already in the base material. Joining by diffusion brazing depends on interdiffusion across the interface to form a liquid layer that wets the base materials (as in the case of the braze alloys consisting of foils). The reaction and interdiffusion of the filler and base materials should also promote the solidification of the liquid. Transient Liquid Phase (TLP) bonding is very similar to diffusion brazing, but solidification occurs due to the formation of high melting temperature phases that are solid at the bonding temperature. Butts (2005) has tested the TLP bonding of Ti-48Al-2Cr-2Nb at.% and a γ–met (Ti-45Al-Nb, B, C at.%) alloys using gas atomised powders of the same composition as the base alloy and copper as a liquid forming component. During brazing, copper melted and dissolved some of the base material powders and isothermal solidification was attained after 10 min at 1,150ºC. PBHT was necessary to eliminate copper-rich regions at the interface. After 60 min at 1,350ºC the bond line became almost microstructurally continuous with the substrates. For the TLP bonding of Ti-48Al-2Cr-2Nb at.% alloy (Gale et al. 2002), the authors used a ratio of 6:1 between alloy powders and copper powders. However, for the γ–met alloy a 50:1 ratio was needed to eliminate concentration gradients at the interface. After the PBHT, the tensile strength of the bonds of the γ–met alloy was similar to that of the base materials after a similar heat treatment, 570 MPa.

The literature cites several works on the application of brazing for joining γ-TiAl (Lee et al. 1998, Lee and Wu 1999, Guedes et al. 2002, 2003, 2004, 2006, Shiue et al. 2004, Wallis et al. 2004, He et al. 2009). Silver and titanium-based filler alloys are those most studied. Titanium-based filler alloys are more interesting for high temperature applications because they produce good quality bonds, forming high melting temperature intermetallic phases, with high mechanical properties, at the interface (Humpston and Jacobson 1993, Cao et al. 2014).

The Ti-Cu-Ni alloy has been successfully used as brazing filler material (Lee et al. 1998, Lee and Wu 1999, Guedes et al. 2003, Wallis et al. 2004). For instance, the investigation of Lee et al. (Lee et al. 1998, Lee and Wu 1999) showed that Ti-15Cu-15Ni (wt.%) can be successfully used for infrared brazing of γ-TiAl, at a temperature range of 1,100 to 1,200°C for 30 to 60 s. For all bonding conditions, the interface presented a multilayered structure. The authors identified different phases at the interface: α(Ti), α_2-Ti_3Al (1,150ºC for 60 s) or α+β (1,150ºC for 30 s), β(Ti) and a residual filler zone at the center of the interface. Further research was reported in the literature on the use of TiCuNi filler materials in the diffusion brazing of these alloys. Guedes et al. (2003) performed the brazing of γ-TiAl in a vacuum furnace with TiCuNi (Ti/Ni,Cu/Ti clad laminated foils) at lower temperatures (980 to 1,000ºC) for longer dwell times (10 min). The microstructural characterisation of the joint interface also revealed a multilayered interface (Figure 2.3); for lower brazing temperature, the interface was characterised by high porosity and unbonded areas. The phase's identification revealed the presence of α_2-Ti_3Al, Ti-Cu-Ni-Al and Ti-Ni-Cu intermetallic compounds at the interface. The hardness of all layers of the interface was higher than the hardness of the base material.

Guedes et al. (2004, 2006) reported other studies that confirmed the successful bonding of Ti-47Al-2Cr-2Nb at.% with TiNi 67 (Ti-33Ni wt.%) filler alloy (Ti/Ni/Ti clad laminated foils). The diffusion brazing experiments were performed in a vacuum furnace at 1,050 and 1,150ºC for 10 min. The interfaces presented two different layers consisting of α_2-Ti_3Al and AlNiTi intermetallics. To improve the mechanical properties of the joints, PBHT were performed. These resulted in microstructural changes

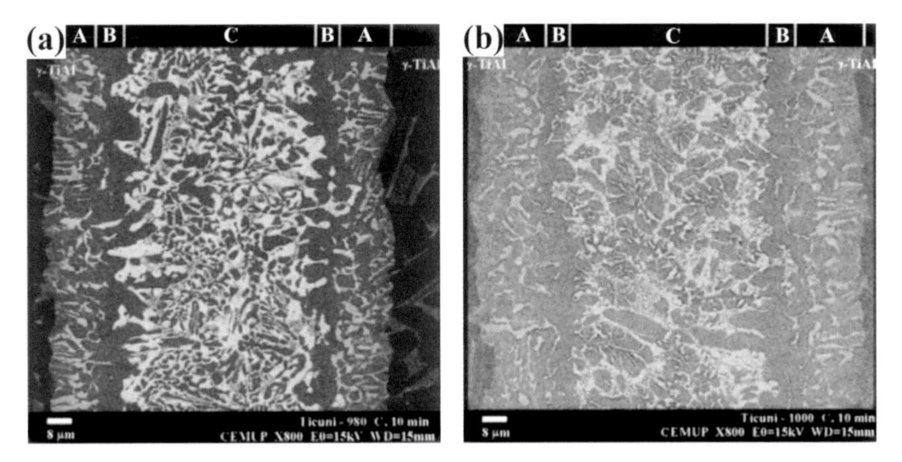

Figure 2.3 SEM images of the γ-TiAl brazing interfaces obtained at: (a) 980°C and (b) 1,000°C. (Reprinted from J. Mater. Sci. 38: 2409–2414. Guedes, A., A.M.P. Pinto, M.F. Vieira and F. Viana, Joining Ti-47Al-2Cr-2Nb with a Ti/(Cu,Ni)/Ti clad-laminated brazed alloy, Copyright 2003, with permission from Springer.)

(layers of α_2-Ti_3Al and AlNiTi transformed into a mixture of γ-TiAl and α_2-Ti_3Al), which led to a decrease in the hardness of the interface. The successful brazing in a vacuum furnace of γ-TiAl alloys was also performed using powders of TiH_2-50 wt.% Ni as filler material, as can be confirmed in the work of He et al. (2009). Sound brazing joints with high shear strength (256 MPa) were obtained at 1,140°C for 15 min. The microstructure of the joints is characterised by a zone comprising $TiAl_3$, Ni_4Ti_3 and α_2-Ti_3Al and a lamellar layer of α_2-Ti_3Al and γ-TiAl phases.

Duan et al. (2004) analysed the possibility of joining γ-TiAl by TLP bonding using a titanium foil combined with copper, nickel or iron foils, where the copper, nickel and iron foils were used as the melting point depressants of titanium. The thicknesses of the foils were 50, 20, 10 and 25 μm, for titanium, copper, nickel and iron respectively. The bond of the base material is promoted by localised melting of the eutectic resulting from the reaction between copper and titanium, nickel and iron. Microstructural characterisation revealed a bonded interface free of defects. However, for all conditions used, α_2-Ti_3Al continuous regions were identified as being normally associated with regions of high brittleness. PBHT promoted a slight improvement in the hardness of the interface due to the low diffusion velocity of copper, nickel and iron in the γ-TiAl alloy. Recently, Lin et al. (2013) demonstrated the application of titanium and nickel as insert materials of γ-TiAl alloys in TLP. The bonding temperatures range from 970 to 1,150°C, with a dwell time of 5 to 240 min. The microstructural observations exhibited defect-free joints. With an increase in bonding temperature, the isothermal solidification rate increased. For temperatures below 1,125°C a α_2-Ti_3Al layer was identified. At 1,150°C a more uniform microstructure was observed, even with a dwell time of 5 min and the joint exhibited the highest shear strength of 281 MPa.

Shiue et al. (2014) used pure silver for infrared brazing γ-TiAl (Ti-50Al, at.%) at temperatures ranging from 1,000 to 1,100°C and dwell times of 15 to 180 min. The ductility of silver-based alloys can be helpful in the relief of residual stress typically produced during cooling from the brazing temperature. The most commonly used Ag-based filler alloys are the eutectic Ag-28Cu (wt.%) due to their excellent wetting properties and their low melting temperature. Guedes et al. (2002) investigated an Ag-based filler with a lower melting temperature, Incusil-ABA (Ag-27.5Cu-12.5In-1.25Ti, wt.%), in order to join a Ti-47Al-2Cr-2Nb at.% alloy in a vacuum furnace at 750°C for 10 min. The bonding interface consisted of an $AlCu_2Ti$ phase formed near the base intermetallic alloy and of a silver solid solution located at its center, in which $AlCu_2Ti$ compounds were dispersed.

The major drawbacks of using Ag-based braze alloys to join γ-TiAl are the lower mechanical properties of the joints in comparison with the ones produced with Ti-based braze alloys. Besides, the joints cannot be used in high temperature applications since these filler alloys exhibit low creep resistance above 400°C (Humpston and Jacobson 1993).

A summary of the brazing conditions, filler materials used for bond γ-TiAl alloys and mechanical properties of the joints can be observed in Table 2.2. The filler alloy has a significant effect on the mechanical properties

Table 2.2 Brazing conditions, filler materials and shear strength values of γ-TiAl joints.

γ-TiAl alloys composition (at.%)	**Filler material**	**Brazing conditions**	**Shear strength (MPa)**
Ti-48Al-2Nb-2Cr (Lee and Wu 1999)	Ti-15Cu-15Ni	1,100–1,200°C/ 30–60 s	319–322
Ti-46Al-4(Cr,Nb,B) (Wallis et al. 2004)	Ti-15Cu-15Ni	1,040°C/600 s	220–230
Ti-43Al-9V-0.3Y (He et al. 2009)	TiH_2+Ni powders	1,180°C Milling time 120 min	256
Ti-45Al-5Nb (Song et al. 2012a)	Ti-38Ni-31Nb	1,220°C/600 s	308
Ti-42.5Al-9V-0.3Y (Song et al. 2012b)	Ti-42Ni-24V	1,220°C/600 s	196
Ti-48Al-2Cr-2Nb (Mirski and Rozanski 2013)	Ag-28Cu	900°C/60 s	149
Ti-48Al-2Cr-2Nb (Lin et al. 2013)	Ti-16.5Ni/Ti-9Ni	1,150°C/300 s	281

of the γ-TiAl joints. This can be explained by the microstructure and phase formation produced by the reaction between the base material and filler material. Brazing at low temperatures can produce an interface free from defects, but with low mechanical properties. Increasing the brazing temperature promotes an increase in mechanical properties.

Based on all the studies reported in the literature it can be concluded that brazing promoted the production of joints free from defects. However, not only is a careful selection of the filler alloy composition essential, but also of the process conditions required. Nonetheless, the formation of a multilayered interface consisting of different phases is characteristic of this process. Usually, these phases are intermetallic compounds that compromise the mechanical properties of the joint. PBHT can lead to microstructural changes, which may be beneficial to the mechanical properties (Guedes et al. 2004, Song et al. 2012a,b). However, the high temperatures and long dwell times required for these treatments significantly increase the costs.

2.5 Concluding Remarks

The development of effective bonding technologies for γ-TiAl alloys is critical for their application in various industrial areas. Conventional fusion welding processes are ineffective for obtaining a crack-free joint, with high mechanical properties, due to their reactivity, the high cooling rates and the formation of residual stresses. An improvement of the mechanical properties of γ-TiAl joints can be achieved by optimisation or modification of the welding processes and by PBHT. γ-TiAl alloys can be successfully joined through other processes such as friction welding and brazing. In general, their mechanical properties are higher than those obtained by fusion welding processes. The interfaces obtained by brazing, diffusion brazing or transient liquid phase bonding processes are characterised by multilayered structures that compromise these mechanical properties. As for welding processes, an improvement in mechanical properties can be achieved by PBHT. However, the high temperatures and long dwell times involved significantly increase the costs of these processes and make their industrial application unfeasible. Diffusion bonding, explained in detail in the following chapters, is an interesting alternative process, since it has some advantages over the processes covered in this chapter.

Keywords: Arc welding; brazing; diffusion bonding; electron beam welding; friction stir welding; friction welding; fusion welding; joining techniques; laser welding; microstructure; mechanical properties; post-bond heat treatment; shear strength; transient liquid phase bonding; tensile strength.

2.6 References

Appel, F., J.D.H. Paul and M. Oehring. 2011. Gamma Titanium Aluminide Alloys: Science and Technology. Wiley-VCH Verlag GmbH & Co. KGaA, Weinheim, Germany.
Arenas, M.F. and V.L. Acoff. 2002. The effect of postweld heat treatment on gas tungsten arc welded gamma titanium aluminide. Scr. Mater. 46: 241–246.
Arenas, M.F. and V.L. Acoff. 2003. Analysis of gamma titanium aluminide welds produced by gas tungsten arc welding. Weld. Res. 110-S–115-S.
Butts, D.A. 2005. Transient liquid phase bonding of a third generation gamma-titanium aluminum alloy- Gamma Met PX. Ph.D. Thesis, Auburn University, Auburn, USA.
Cao, J., J. Qi, X. Song and J. Feng. 2014. Welding and joining of titanium aluminides. Materials 7: 4930–4962.
Chaturvedi, M.C., N.L. Richards and Q. Xu. 1997. Electron beam welding of a Ti-45Al-2Nb-2Mn +0.8 vol.% TiB_2 XD alloy. Mater. Sci. Eng. A-Struct. Mater. Prop. Microstruct. Process. 239-240: 605–612.
Chaturvedi, M.C., Q. Xu and N.L. Richards. 2001. Development of crack-free welds in a TiAl-based alloy. J. Mater. Process. Technol. 118: 74–78.
Chen, G., B. Zhang, W. Liu and J. Feng. 2011. Crack formation and control upon the electron beam welding of TiAl-based alloys. Intermetallics 19: 1857–1863.
Dong, H., L. Yu, D. Deng, W. Zhou and C. Dong. 2015. Direct friction welding of TiAl alloy to 42CrMo steel rods. Mater. Manuf. Process. 30: 1104–1108.
Duan, H., M. Koçak, K.-H. Bohm and V. Ventzke. 2004. Transient liquid phase (TLP) bonding of TiAl using various insert foils. Sci. Technol. Weld. Joi. 9: 513–518.
Gale, W.F., D.A. Butts, M. Di Ruscio and T. Zhou. 2002. Microstructure and mechanical properties of titanium aluminide wide-gap, transient liquid-phase bonds prepared using a slurry-deposited composite interlayer. Metall. Mater. Trans. A-Phys. Metall. Mater. Sci. 33A: 3205–3214.
Guedes, A., A.M.P. Pinto, M.F. Vieira, F. Viana, A.S. Ramos and M.T. Vieira. 2002. Microstructural characterisation of γ-TiAl joints. Key Eng. Mater. 230-232: 27–30.
Guedes, A., A.M.P. Pinto, M.F. Vieira and F. Viana. 2003. Joining Ti-47Al-2Cr-2Nb with a Ti/(Cu,Ni)/Ti clad-laminated brazed alloy. J. Mater. Sci. 38: 2409–2414.
Guedes, A., A.M.P. Pinto, M.F. Vieira and F. Viana. 2004. Joining Ti-47Al-2Cr-2Nb with a Ti-Ni braze alloy. Mater. Sci. Forum 455-456: 880–884.
Guedes, A., A.M.P. Pinto, M.F. Vieira and F. Viana. 2006. Assessing the influence of heat treatments on γ-TiAl joints. Mater. Sci. Forum 514-516: 1333–1337.
He, P., J.C. Feng and W. Xu. 2005. Interfacial microstructure of induction brazed joints of TiAl-based intermetallics to steel 35CrMo with AgCuNiLi filler. Mater. Sci. Eng. A-Struct. Mater. Prop. Microstruct. Process. 408: 195–201.
He, P., D. Liu, E. Shang and M. Wang. 2009. Effect of mechanical milling on Ni–TiH2 powder alloy filler metal for brazing TiAl intermetallic alloy: The microstructure and joint's properties. Mater. Charact. 60: 30–35.
Hirose, A., Y. Arita and K.F. Kobayashi. 1995. Microstructure and crack sensitivity of laser-fusion zones of Ti-46 mol% Al-2 mol% Mo alloy. J. Mater. Sci. 30: 970–979.
Humpston, G. and D.M. Jacobson. 1993. Principles of Soldering and Brazing. ASM International Metals Park, Ohio, USA.
Kestler, H. and H. Clemens. 2003. Production, processing and application of γ(TiAl)-based alloys. pp. 351–388. *In:* C. Leyens and M. Peters [eds.]. Titanium and Titanium Alloys: Fundamentals and Applications. Wiley-VCH Verlag GmbH & Co. KGaA, Weinheim, Germany.
Lee, S.J., S.K. Wu and R.Y. Lin. 1998. Infrared joining of TiAl intermetallics using Ti-15Cu-15Ni foil-1. The microstructure morphologies of joint interfaces. Acta Mater. 46: 1283–1295.
Lee, S.J. and S.K. Wu. 1999. Infrared joining strength and interfacial microstructures of Ti-48Al-2Nb-2Cr intermetallics using Ti-15Cu-15Ni foil. Intermetallics 7: 11–21.

Lee, W.-B., M.-G. Kim, J.-M. Koo, K.-K. Kim, D.J. Quesnel, Y.-J. Kim et al. 2004a. Friction welding of TiAl and AISI 4140. J. Mater. Sci. 39: 1125–1128.
Lee, W.-B., Y.-J. Kim and S.-B. Jung. 2004b. Effects of copper insert layer on the properties of friction welded joints between TiAl and AISI 4140 structural steel. Intermetallics 12: 671–678.
Lin, T., H. Li, P. He, H. Wei, L. Li and J. Feng. 2013. Microstructure evolution and mechanical properties of transient liquid phase (TLP) bonded joints of TiAl intermetalllics. Intermetallics 37: 59–64.
Liu, J., V. Ventzke, P. Staron, N. Schell, N. Kashaev and N. Huber. 2012. Investigation of *in situ* and conventional post-weld heat treatments on dual-laser-beam-welded γ-TiAl-based alloy. Adv. Eng. Mater. 14: 923–927.
Liu, J., V. Ventzke, P. Staron, N. Schell, N. Kashaev and N. Huber. 2014. Effect of post-weld heat treatment on microstructure and mechanical properties of laser beam welded TiAl-based alloys. Metall. Mater. Trans. A-Phys. Metall. Mater. Sci. 45A: 16–28.
Lütjering, G. and J.C. Williams. 2007. Titanium. 2nd edition. Springer-Verlag Berlin, Heidelberg, Germany.
Mallory, L.C., W.A. Baeslack and D. Phillips. 1994. Evolution of the weld heat-affected zone microstructure in a Ti-48Al-2Cr-2Nb gamma titanium aluminide. J. Mater. Sci. Lett. 13: 1061–1065.
Messler Jr., R.W. 2004a. Principles of Welding Processes, Physics, Chemistry, and Metallurgy. Wiley-VCH Verlag GmbH & Co. KGaA, Weinheim, Germany.
Messler Jr., R.W. 2004b. Joining of Materials and Structures: from Pragmatic Process to Enabling Technology. Elsevier Butterworth-Heinemann, Burlington, MA, USA.
Mirski, Z. and M. Rozanski. 2013. Diffusion brazing of titanium aluminide alloy based on TiAl (γ). Arch. Civ. Mech. Eng. 13: 415–421.
Miyashita, T. and H. Hino. 1994. Friction welding characteristics of TiAl intermetallic compound. J. Japan. Inst. Metals 58: 215–220.
Nakao, Y., K. Shinozaki and M. Hamada. 1991. Diffusion bonding of intermetallic compound TiAl. ISIJ Int. 31: 1260–1266.
Ranatowski, E. 2008. Weldability of titanium and its alloys-progress in joining. Adv. Mater. Sci. 8: 69–76.
Shiue, R.K., S.K. Wu and S.Y. Chen. 2004. Infrared brazing of TiAl intermetallic using pure silver. Intermetallics 12: 929–936.
Song, X.G., J. Cao, Y.Z. Liu and J.C. Feng. 2012a. Brazing high Nb containing TiAl alloy using TiNi-Nb eutectic braze alloy. Intermetallics 22: 136–141.
Song, X.G., J. Cao, H.Y. Chen, Y.F. Wang and J.C. Feng. 2012b. Brazing TiAl intermetallics using TiNi-V eutectic brazing alloy. Mater. Sci. Eng. A-Struct. Mater. Prop. Microstruct. Process. 551: 133–139.
Sun, J.G., A.J. Kropf, D.R. Vissers, W.M. Sun, J. Katsoudas, N. Yang et al. 2012. Synchrotron X-ray CT characterization of friction-welded joints in TiAl turbocharger components. AIP Conf. Proc. 1430: 1251–1258.
Threadgill, P.L. 1995. The prospects for joining aluminides. Mater. Sci. Eng. A-Struct. Mater. Prop. Microstruct. Process 192–193: 640–646.
Ventzke, V., H.-G. Brokmeier, P. Merhof and M. Koçak. 2010. Microstructural characterization of friction welded TiAl-Ti6Al4V hybrid joints. Solid State Phenom. 160: 319–326.
Wallis, I.C., H.S. Ubhi, M.-P. Bacos, P. Josso, J. Lindqvist, D. Lundstrom et al. 2004. Brazed joints in γ TiAl sheet: Microstructure and properties. Intermetallics 12: 303–316.
Xu, Q., M.C. Chaturvedi and N.L. Richards. 1999. The role of phase transformation in electron beam welding of TiAl-based alloys. Metall. Mater. Trans. A-Phys. Metall. Mater. Sci. 30A: 1717–1726.

CHAPTER 3

Diffusion Bonding of γ-TiAl Alloys

3.1 Introduction

Diffusion bonding is a solid-state process that occurs through the interdiffusion of atoms across the joining interface. The formation of the bonds results from the closure of the mating surfaces due to applied pressure at an elevated temperature that promotes interdiffusion at the surface layers of the materials being joined.

The joints are characterised by uniform microstructures and mechanical properties throughout the interface, without discontinuities, oxide inclusions, voids or loss of alloying elements. Diffusion bonding is appropriate for joining sensitive-temperature materials, for the production of components with complex shapes and near-net shape joints with accurate dimensions, due to the minimum distortion and deformation associated with this process. It can be used in similar and dissimilar bonding, especially between materials with different thermo-physical properties. Another interesting aspect is that it can be combined with other processes, such as superplastic forming.

However, diffusion bonding requires high temperature, pressure and a long dwell time in a vacuum atmosphere, which limits the feasibility of this process for the production of large series. The initial investment in an equipment is significant and the production of large components is limited by the size of the equipment used.

This process is most promising in the joining of γ-TiAl alloys, since it avoids the metallurgical problems associated with the fusion-welding processes, leading to the formation of sound joints. The interfaces are very thin, more than those produced by processes such as brazing, presenting mechanical properties higher than or similar to the ones of the base materials.

The following sections will briefly describe the diffusion bonding process and its applicability for joining γ-TiAl alloys. The diffusion mechanisms in intermetallic compounds will also be discussed, especially for the γ-TiAl alloys.

3.2 Diffusion Bonding

Diffusion bonding technology is essentially controlled by diffusion mechanisms. For the diffusion to occur and form a joint free of porosity, temperatures higher than 50 per cent of the absolute melting temperature of the base materials are used. A careful surface preparation is essential for the removal of oxides and contaminants, assuring the contact of the mating surfaces at the atomic level during bonding (Kestler and Clemens 2003, Messler Jr. 2004, Lütjering and Williams 2007, Zhou 2008).

However, even with careful surface preparation of the base materials, they still exhibit some roughness and oxide layers. For this reason, it is essential to apply pressure during the initial stage of bonding in order to promote plastic deformation of the mating surfaces and maintain contact, while keeping the pressure low to avoid macroscopic deformation of the base materials. The time at the bonding temperature can range from a few minutes to several hours, depending on the applied pressure and temperature and also on the materials to be joined (Messler Jr. 2004, Zhou 2008).

Diffusion bonding can be divided into three stages: plastic deformation at the initial stage, grain boundary diffusion, and lastly, volume diffusion and pore elimination (Lütjering and Williams 2007, Zhou 2008).

At the first stage, the micro asperities of the bonding surfaces ensure the initial contact of the base materials. The applied pressure causes local high stresses (the yield strength of the material is exceeded at the contact points) that results in the breakdown of surface oxide layers and in plastic deformation, which increases the contact area and creates a flat interface. At the end of the first step, the bonded area is less than 10 per cent and a large number of pores and oxides remain between the surfaces to be joined. With the increase in the bonding temperature, thermal activated mechanisms such as diffusion and creep begin to occur, resulting in an increase in the local plastic deformation. At the second stage, diffusion continues to increase and becomes dominant. The joining process is then controlled by migration of the grain boundaries. Some pores are eliminated and the contact area becomes discontinuous. Some of the pores, which were not previously eliminated, are retained within the grains, thus making their removal more difficult. The last stage is volume diffusion, which eliminates or reduces the pores at the interface (Messler Jr. 2004, Lütjering and Williams 2007, Zhou 2008).

A schematic illustration of the equipment for solid-state diffusion bonding can be observed in Figure 3.1. The diffusion bonding equipment includes a furnace, a press (uniaxial or isostatic) and a vacuum chamber.

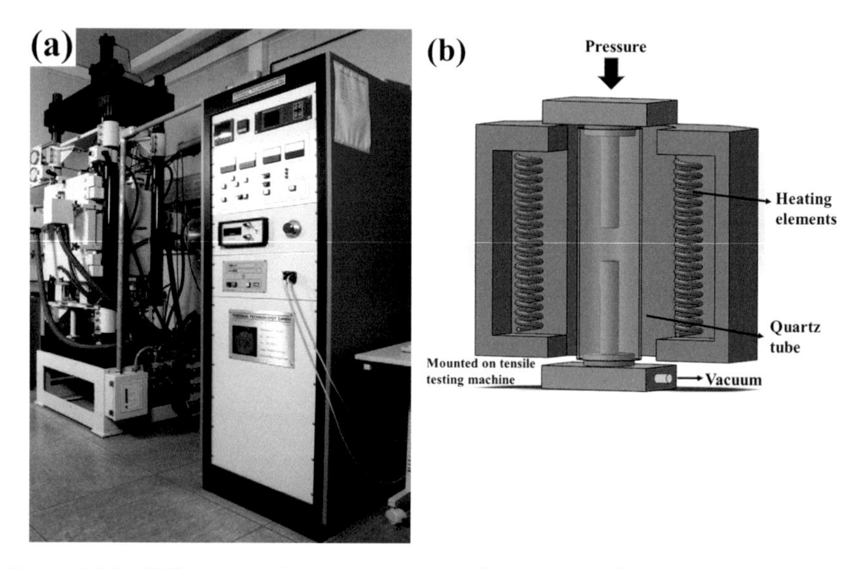

Figure 3.1 (a) Diffusion bonding equipment and (b) schematic illustration of the furnace, uniaxial press and vacuum chamber.

3.2.1 Diffusion Bonding Parameters

In general, there are five important joining parameters in the diffusion bonding process: the quality of the mating surfaces, temperature, dwell time, pressure and vacuum level. Precise control of these parameters is important in order to obtain sound joints and ensure diffusion across the interface so as to produce a uniform microstructure and high mechanical properties (Messler Jr. 2004, Zhou 2008). The influence of each of these variables in the process will be discussed below.

The quality of the mating surfaces is one of the parameters that most significantly affects the joining. A good initial contact between the surfaces of the base materials is very important for this process. For an intimate contact, it is necessary that the mating surfaces meet two critical requirements: they should be flat and free from contamination and oxide layers.

The average roughness values recommended for this type of bonding range between 0.4 and 0.8 μm (Messler Jr. 2004). The presence of oxide layers on the contact surfaces may induce poor bonding, in particular, if they are stable and do not decompose or dissolve at the bonding temperature. The adoption of isostatic pressure significantly reduces the quality of surface preparation required for a successful joint.

Among the major parameters of the process, the temperature is by far the most important, assuming there is sufficient pressure to ensure initial contact between the surfaces of the base materials. The temperature reduces the yield strength of the base materials, facilitating the deformation of the surface asperities and contact of the surfaces. Apart from this effect, the main role of the temperature in the process is to facilitate atomic diffusion, thereby creating the bond. The atomic flux of atoms across the interface is proportional to the diffusion coefficient whose variation with temperature obeys an Arrhenius equation—grows exponentially with the temperature. A bonding temperature of 50 to 80 per cent of the absolute melting temperature of the base materials is essential in order to assure a high diffusion rate.

The bonding time can range from a few minutes to several hours. The diffusion distances increase with an increase in the bonding time, which leads to the formation of joints with less porosity and improved mechanical properties. However, it is necessary to reach a commitment, since longer bonding times may also cause microstructural changes, such as grain growth (Kestler and Clemens 2003, Messler Jr. 2004, Lütjering and Williams 2007, Zhou 2008).

As already mentioned, pressure is also an important factor in the solid-state diffusion bonding process, and its influence depends on the process stage. At the first stage, pressure is critical for establishing contact between the mating surfaces, to break the oxide layers and to allow subsequent diffusion and recrystallization. The applied pressure causes high local stresses, due to the reduced area of contact, and the yield strength of the base materials is exceeded.

The plastic deformation of the protrusions of the surface increases the contact area and causes rupture of the surface layers of oxides and contaminants. In the following stages, high pressures can cause a negative effect as they can promote microstructural changes (Messler Jr. 2004). As already mentioned, it is important to keep the pressure low to avoid macroscopic deformation of the base materials.

The vacuum level is a variable that depends on the base materials and also on the bonding conditions. The use of a vacuum atmosphere is very important in diffusion bonding at high temperatures in order to inhibit the oxidation of the base materials and to eliminate contaminations. An important issue is the determination of the level of vacuum required to obtain a good balance between the quality of the resulting joints and the associated costs. This demand will not only depend on the materials to be joined, but also on the upcoming applications. Vacuum levels of 100 to 1 Pa are usual in diffusion bonding (Kestler and Clemens 2003, Messler Jr. 2004, Lütjering and Williams 2007, Zhou 2008).

In addition to these process parameters, which must be carefully selected, other variables related to metallurgical factors, including allotropic transformation and the occurrence of recrystallization, should be taken into consideration. These features can modify the diffusion rates, thereby affecting the diffusion bonding process (Kestler and Clemens 2003, Messler Jr. 2004, Lütjering and Williams 2007, Zhou 2008).

3.2.2 Interdiffusion in γ-TiAl Intermetallics

The knowledge of diffusion characteristics and fundamental understanding of diffusion mechanisms are of great importance for the research and design of diffusion bonding of γ-TiAl alloys. Due to the ordered structure of intermetallic phases, diffusion mechanisms are more complex and involve sequences of jumps of several atoms. The crystal order is destroyed, only locally and temporarily, but is totally restored when the sequence is complete (Divinski and Herzig 2000, Mehrer 2007).

Herzig et al. (1999) described experimental and theoretical studies of titanium and aluminum self-diffusion in γ-TiAl. The authors determined the most favourable mechanism for self-diffusion in γ-TiAl: the nearest neighbour jumps through vacancies in the titanium sublattice, by the sublattice diffusion mechanism. Random jumps of the vacancies on this sublattice will not affect the order of the compound and will provide long-range diffusion.

At high temperatures, $T > 1{,}450$ K, the anti-structure bridge mechanism makes a major contribution to the diffusion of titanium, causing a deviation from the Arrhenius temperature dependence (Herzig et al. 1999). The vacancy concentration on the aluminum sublattice is generally much smaller, up to several orders of magnitude, than on the titanium sublattice (Herzig et al. 1999, Mehrer 2007), which makes diffusion difficult. Diffusion of aluminum atoms occurs by means of the jump to the titanium sublattice, diffusion within this sublattice and then the jump back to the aluminum side. Aluminum diffusion follows a perfect Arrhenius temperature dependence.

The γ-TiAl structure can be seen as alternated (002) layers of titanium and aluminum. Thus, the diffusion of titanium atoms is faster along the layers than in the perpendicular [001] direction. However, the diffusion anisotropy is attenuated at high temperatures due to increased diffusion through the anti-structure bridge mechanism. According to Nakajima et al. (1997), the interdiffusivity in γ-TiAl shows no concentration dependence between 50–54 per cent aluminum.

The Arrhenius parameters (D_0 and Q) for titanium and aluminum self-diffusion in γ-TiAl are presented in Table 3.1 (Lütjering and Williams 2007).

Table 3.1 The Arrhenius parameters for self-diffusion in γ-TiAl (Lütjering and Williams 2007).

Diffuser	D_0 (m^2s^{-1})	Q (eV)
Al	2.11×10^{-2}	3.71
Ti	1.43×10^{-6}	2.59

3.3 Diffusion Bonding of γ-TiAl Alloys

Over the last twenty-five years, several studies have been reported in the literature about the solid-state diffusion bonding of γ-TiAl alloys (Nakao et al. 1991, Yan and Wallach 1993, Godfrey et al. 1993, Glatz and Clemens 1997, Holmquist et al. 1998, Çam and Koçak 1999, Çam et al. 1999, 2006, Zhao et al. 2006, Hermann and Appel 2009). These have demonstrated the applicability of diffusion bonding in these alloys by the production of joints with the desired microstructure and mechanical properties.

Nakao et al. (1991) studied the solid-state diffusion bonding of the Ti-52Al at.% alloy. For this purpose, the experiments were carried out in a vacuum at temperatures of 1,000, 1,100 and 1,200°C, at dwell times of between 16 and 64 min and pressures varying from 10 to 30 MPa. Joints produced at 1,200°C for 15 min and under a pressure of 10 MPa exhibited pores and oxides such as TiO_2 and Al_2TiO_5 at the interface. These defects disappeared at longer bonding times and at a higher pressure. Sound joints without defects were obtained at 1,200°C for 64 min under a pressure of 15 MPa. The mechanical properties of these joints were evaluated by tensile tests at room temperature and high temperatures (800 and 1,200°C). At room temperature the tensile strength was 225 MPa and the fracture occurred, for most of the tensile specimens, through the base material thereby confirming the strength of the joints. At high temperatures, 800 and 1,200°C, the tensile strength was lower than that of the γ-TiAl alloy, the fracture occurring at the bonding interface. The mechanical properties of the joints at high temperature were improved by a Post-Bond Heat Treatment (PBHT) at 1,300°C. This treatment, which promoted grain growth more intensively at the bonded zone, increased the tensile strength to 210 MPa at 1,000°C with the fracture occurring at the base material.

Yan and Wallach (1993) studied the diffusion bonding of γ-TiAl alloy with 48 at.% aluminum. Sound joints were produced by solid-state diffusion bonding conducted at temperatures of at least 1,100°C. The authors selected the surface roughness and the bonding pressure in such a way that the minimum plastic strain needed to activate the dynamic recrystallization was achieved at the interface. In this case, the bond line disappeared and the microstructure of the interface revealed fine recrystallized γ-TiAl grains. The higher shear strength was determined for samples processed at 1,200°C under a pressure of 20 MPa; a value of 260 MPa was observed that was close to the value of the base material (300 MPa).

Godfrey et al. (1993) investigated the influence of temperature (range from 1,200 to 1,350°C), dwell time (15 and 45 min) and microstructure of the base material (fully lamellar, duplex and fully γ) in the diffusion bonding of the Ti-48Al-2Mn-2Nb at.% alloy. A constant pressure of 10 MPa was used in all experiments. The microstructure of the interface was composed mainly of twin-related γ-TiAl grains with a size of about 10 to 20 μm. Due to the high temperatures used during joining, grain growth and changes in the microstructure of the base material, especially for the lamellar microstructure, were reported.

Glatz and Clemens (1997) performed solid-state diffusion bonding of Ti-47Al-2Cr-0.2Si at.% alloy. Joints free from defects were produced at 1,000°C, for dwell times of 60 and 180 min and pressures ranging from 20 to 40 MPa. The processing conditions introduced no relevant changes in the fine-grained microstructure of the base material. The mechanical properties of these joints were evaluated by tensile tests at room temperature and at high temperature (700 and 1,000°C). The tensile strength at room temperature was similar to that exhibited by the base material (600 MPa) and fracture occurred in the base material. For temperatures of 700 and 1,000°C, the mechanical strength was also similar to the base material, but the fracture occurred in the interface with a significantly lower elongation in some cases.

Çam et al. (Çam and Koçak 1999, Çam et al. 1999, 2006) studied the solid-state diffusion bonding of several γ-TiAl alloys. Early studies (Çam and Koçak 1999, Çam et al. 1999) focused on diffusion bonding of the Ti-47Al-4.5(Cr, Mn, Nb, Si, B) at.% alloy. Joining experiments were performed at temperatures ranging from 950 to 1,100°C, with pressures of 20, 30 and 40 MPa and bonding times of 60 and 180 min (Çam and Koçak 1999). The joints produced at 950 and 1,000°C under a pressure of 30 MPa were defect-free, but the bond lines were clearly visible.

The roughness of the mating surfaces and the local deformation provoked by the bonding pressure promoted dynamic recrystallization with the formation of a thin layer of fine γ grains at the interface. This visible bond line was a preferential crack path, explaining the lower shear strength of these joints and the occurrence of a fracture predominantly at the bond interface. The change in processing conditions (for example by increasing the temperature to 1,100°C) improved diffusion and thus the grain growth kinetics in this region. γ grains grew significantly, the bond line was no longer visible and the shear strength increased, with the fracture occurring through the base material.

A PBHT performed at 1,430°C for 30 min improved the joint quality and the shear strength increased, reaching a maximum value of 466 MPa, very similar to that of the base material. However, this treatment modified the base material's microstructure. In another study, the authors

(Çam et al. 1999) optimised the solid-state diffusion bonding processing conditions of these γ-TiAl alloys using the temperature of 1,000°C (one of the lowest temperatures), lower pressures (5, 10 and 20 MPa) and longer bonding times (300 and 480 min). The joints produced at bonding times of 300 min and pressures of 10 and 20 MPa or at 480 min and 5 MPa led to the disappearance of the bond line. The joints processed over longer times and at higher pressures (480 min and 20 MPa) exhibited the highest shear strength value (542 MPa).

Çam et al. (2006) also investigated the diffusion bonding of a γ-TiAl alloy containing carbon: Ti-47Al-3.7(Nb, Cr, C) at.%. The joining experiments were performed at 950 and 1,000°C, with bonding times of 60, 180 and 300 min and pressures of 5, 10 and 20 MPa. The results were similar to those obtained using the other alloy (Çam and Koçak 1999, Çam et al. 1999). For all conditions, the bonds were of good quality, without cracks or pores. However, the bond line was visible and the formation of the brittle α_2-Ti_3Al phase was detected at the interface.

Figure 3.2 shows the microstructure of the bonds produced at 1,000°C with different dwell times and pressures. The greatest shear strength values were obtained for the longest dwell time tested at each of the two higher

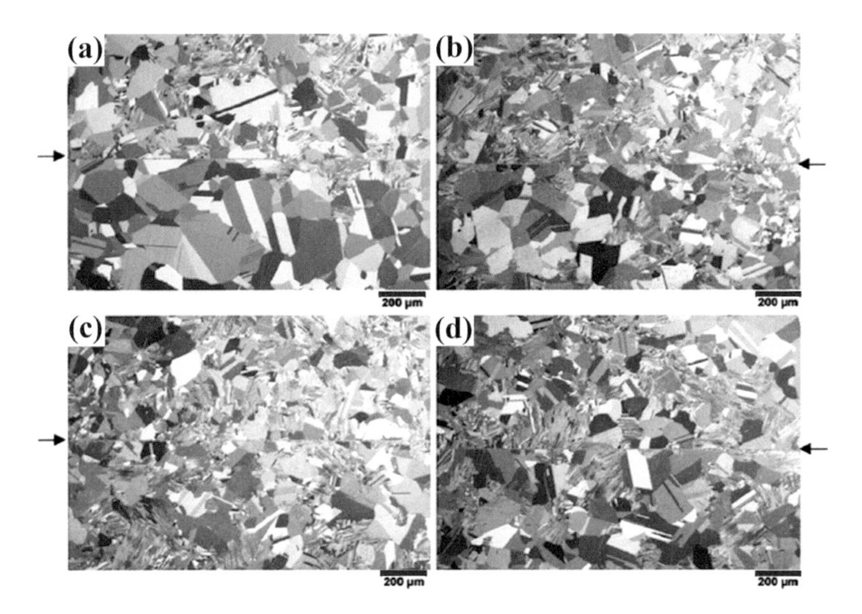

Figure 3.2 Polarised light micrographs of the bonds produced at 1,000°C: (a) under a pressure of 5 MPa for 300 min, (b) under a pressure of 10 MPa for 60 min, (c) under a pressure of 10 MPa for 300 min and (d) under a pressure of 20 MPa for 180 min. (Reprinted from J. Mater. Sci. 41: 5273–5282. Çam, G., G. Ipekoğlu, K.-H. Bohm and M. Koçak. Investigation into the microstructure and mechanical properties of diffusion bonded TiAl alloys. Copyright 2006, with permission from Springer.)

pressures (300 min and 180 min for 10 and 20 MPa, respectively). A PBHT promoted the elimination of the bond line, thus improving shear strength, but significantly changing the microstructure of the base material.

Figure 3.3 shows the microstructure of the bonds after the PBHT at 1,430°C. All the bonds are characterised by a coarse-grained fully lamellar microstructure.

Herrmann and Appel (2009) studied the diffusion bonding of the γ-TiAl alloy with different compositions. Eight γ-TiAl alloys with different microstructures were used: Ti-44.5Al at.% (duplex), Ti-45Al-10Nb at.% (fully lamellar), Ti-46.5Al at.% (duplex, nearly globular), Ti-46.5Al-5.5Nb at.% (nearly globular), Ti-47Al-4.5Nb-0.2C-0.2B at.% (duplex, nearly globular), Ti-45Al-8Nb-0.2C at.% (duplex, nearly globular), Ti-45Al-5Nb-0.2C-0.2B at.% (nearly globular) and Ti-54Al at.% (globular γ).

Bonding experiments were performed at 1,000°C, for 15 to 120 min and under pressures ranging from 20 to 100 MPa. The microstructure of the bonding interface typically showed three distinct zones: a layer composed

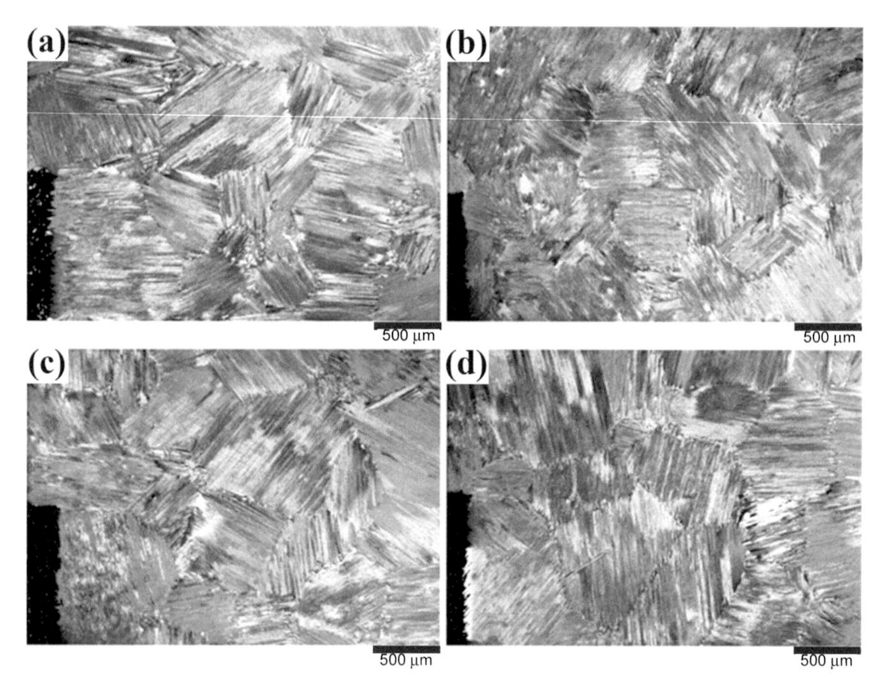

Figure 3.3 Polarised light micrographs after PBHT at 1,430°C for the bond produced at (a) 950°C under a pressure of 10 MPa for 180 min, (b) 1,000°C under a pressure of 10 MPa for 60 min, (c) 1,000°C under a pressure of 10 MPa for 180 min and (d) 1,000°C under a pressure of 10 MPa for 300 min. (Reprinted from J. Mater. Sci. 41: 5273–5282. Çam, G., G. Ipekoğlu, K.-H. Bohm and M. Koçak. Investigation into the microstructure and mechanical properties of diffusion bonded TiAl alloys. Copyright 2006, with permission from Springer.)

of a very fine-grained α_2-Ti_3Al phase, which formed along the contact area between the two joining surfaces (bond line), a region of relatively large dynamically recrystallized γ grains, and a region of deformed and partially recovered grains of the base material. The extent of these regions depended on the bonding conditions as well as on the composition and microstructure of the γ-TiAl alloy. For each alloy, an accurate optimisation of the bonding parameters was necessary in order to avoid large-scale deformation of components, caused by too high pressure, and the presence of pores and voids, developed with short dwell times and low pressures.

The bondability of the various types of alloys was attested in function of metallographic analysis. Niobium-bearing alloys with lower Al content exhibited the best mechanical behaviour. In these alloys, the dynamic recrystallization was facilitated and the formation of α_2-Ti_3Al was avoided. The two-phase $(\gamma + \alpha_2)$ alloys were the second easiest alloys to bond. The dynamic recrystallization was slower than in the previous alloys, but the main disadvantage was the presence of a layer of brittle α_2 phase grains at the bond line, which impairs the mechanical behaviour of the joint. Finally, Al-rich single-phase γ-TiAl alloys were the most difficult to bond because of the greater difficulty in the nucleation and growth of recrystallized grains and the formation of oxides and nitrides at the interface. Tensile tests on Ti-45Al-5Nb-0.2C-0.2B at.% alloys diffusion bonded at 1,000°C for 120 min under a pressure of 20 MPa were performed to evaluate the mechanical strength. At room temperature, the samples fractured by the bond line with a tensile strength of 830 MPa, which is approximately 85 per cent of the yield stress of the base material.

Recently, the Spark Plasma Sintering (SPS) technique has been applied for diffusion bonding a Ti-45Al-7Nb-0.3W at.% alloy (Zhao et al. 2006). The bonding temperatures were 1,000, 1,100, 1,150 and 1,200°C, with bonding times ranging from 15 to 60 min and pressures of 15, 20 or 30 MPa. The strength of the joints was evaluated by tensile tests. The tensile strength increased with bonding time and temperature and exceeded the strength of the γ-TiAl alloy when joined at 1,200°C for 60 min (fracture occurred through the base material). Conversely, increasing the pressure lowered the tensile strength. The authors compared the SPS technique with the conventional hot pressing technique and concluded that SPS required a lower pressure and shorter dwell time to produce good quality joints of this γ-TiAl alloy.

The Superplastic Forming and Diffusion Bonding (SPF/DB) process, which combines a forming operation with diffusion bonding, has been suggested as a possibility for the production of very complex components of γ-TiAl alloys (Imayev et al. 1992, Lutfullin et al. 1995). The forming operation explores these alloys superplastic behaviour, which allows them to be deformed beyond their normal range of plastic deformation. This process, when compared with the conventional techniques, permits a reduction in the cost and weight of components. In fact, and even considering the investment in tooling and modified raw materials, the SPF/DB process allows fabrication of complex products in one operation, thus substantially reducing assembly costs. Superplastic forming can be performed within limited ranges of temperature, strain rate and with proper microstructures. Typically, the exceptional ductility is achieved with a fine and stable grain size (Lutfullin 1995, Xun and Tan 2000, Wu and Huang 2001, 2003, Wu et al. 2004, Hefti 2008). Since the early 90s, the superplastic behaviour of γ-TiAl alloys has been demonstrated. Imayev et al. (1992) have shown that a near-equiatomic γ-TiAl alloy, which had a mean grain size of 0.4 μm, when tested in tension at 800ºC with an initial strain rate of $8.3 \times 10^{-4}\ s^{-1}$, exhibited an elongation to rupture of 225 per cent.

Wu et al. (Wu and Huang 2001, 2003 and Wu et al. 2004) investigated the SPF/DB process of γ-TiAl alloys. Laser surface remelting, followed by PBHT, was conducted to induce a fine-grained structure with a grain size of approximately 1 μm, which was essential for the alloys to exhibit superplastic behaviour. This superplasticity enabled bonding at lower temperatures. The diffusion bonding of a titanium aluminide alloy, Ti–46.5Al–2Cr–1.5Nb–1V at.%, was studied and the results obtained with and without laser surface processing were compared (Wu et al. 2004). For bonding at 900°C for 120 min under a pressure of 60 MPa, the refinement of the microstructure of the surface clearly improved the quality of the joint.

In fact, for these processing conditions, the shear strength of the bond increased from 266 to 384 MPa through the use of the laser surface modification. For conventional diffusion bonding experiments, similar values of shear strength were only obtained for more demanding processing conditions (shear strength of 200 MPa for bonding at 1,100ºC/30 MPa/60 min). However, even with laser treatment and pre-bond and post-bond heat treatments, the maximum shear strength was always lower than that of the base material (486 MPa), with the fracture occurring through the interface. Similar results were obtained with a different alloy: Ti-45Al-2Mn-2Nb-0.8 vol.% TiB_2 (Wu and Huang 2003).

Zhu et al. (2005) also investigated the superplastic diffusion bonding of γ-TiAl alloy. A Ti–48Al–2Cr–2Nb at.% alloy with a coarse-grained fully lamellar microstructure was subjected to a thermomechanical process involving multiple steps of forging. After thermomechanical processing, the alloy showed a fine-grained duplex microstructure with an average grain size of about 4 μm. The fine-grained γ-TiAl alloy had a tensile elongation of 300 per cent when deformed at 1,100ºC and at a strain rate of $8.3 \times 10^{-5}\ s^{-1}$. The diffusion bonding process produced sound joints at 1,100ºC for 60 min under a pressure of 10 MPa.

Summing up, γ-TiAl diffusion bonds can be produced at temperatures ranging from 950 to 1,400°C with bonding times extending from a few minutes to several hours under pressures up to 100 MPa. Diffusion bonding conditions of γ-TiAl alloys depend on the composition, microstructure and processing technique.

Table 3.2 shows the diffusion bonding conditions and associated mechanical properties revealed by the most important studies.

For some applications in the aerospace and automotive industries, the manufacture of multi-material components is essential. In these cases, the joining of dissimilar materials, such as γ-TiAl alloys to conventional materials, is of crucial interest. Situations with greater potential of industrial application are the joining of γ-TiAl to steels, titanium alloys or nickel superalloys. However, as with similar joints, welding processes do not easily lend themselves to the successful joining of γ-TiAl alloys to other materials. Therefore, several studies were conducted to optimise the dissimilar diffusion bonding of these materials.

In this section, the diffusion bonding of γ-TiAl to titanium alloys (especially to the Ti6Al4V alloy) is presented, since the joining of γ-TiAl alloys to steel and to Ni-based superalloys will be described in detail in Chapters 6 and 7.

Glatz and Clemens (1997) presented the first results concerning the joining of dissimilar joints of γ-TiAl alloy (Ti-47Al-2Cr-0.2Si at.%) to titanium alloy (Ti6Al4V). The selected processing parameters were 1,000°C for 180 min under a pressure of 20 MPa. Apart from some isolated micropores, good quality bonds were achieved. The tensile strength values obtained by tests at room temperature and at 700°C were 558 and 227 MPa, respectively. The fracture occurred through the γ-TiAl for the samples tested at room temperature while those tested at 700ºC exhibited a fracture through the titanium alloy. Despite the good results obtained, the temperature and the pressure used were too high for the titanium alloy (Ti6Al4V), leading to its plastic deformation.

Holmquist et al. (1998) studied the dissimilar joining of Ti-33Al-2Fe-1.8V-0.1B (wt.%) to Ti6Al4V using a hot isostatic pressure technique at 900, 940 and 980°C for 60 min under a pressure of 200 MPa. All the bonds produced under these bonding conditions were of good quality and free from defects. A strong change in the microstructure of the titanium alloy was observed for bonds produced at the two higher temperatures.

The mechanical characterisation of the joints was carried out by tensile tests. The joints exhibited a mechanical behaviour similar to the γ-TiAl alloy. However, fracture occurred almost always through the bond line. These authors also studied the bonding of this γ-TiAl alloy to the Ti-5.8Al-4.0Sn-3.5ZN-0.7Nb-0.5Mo-0.35Si-0.06C (wt.%) titanium alloy (Holmquist et al. 1999). The same processing conditions were used in the bonding, but only at the lower temperature (900°C) to avoid microstructural changes to the titanium alloy. As in the previous study, sound joints were produced, the joints presenting a tensile strength similar to γ-TiAl and fracture occurring

Table 3.2 Diffusion bonding conditions of γ-TiAl alloys.

γ-TiAl alloys composition (at.%)	Diffusion bonding parameters			Tensile strength (MPa)	Shear strength (MPa)
	Temperature (°C)	Time (min)	Pressure (MPa)		
Ti-52Al (Nakao et al. 1991)	1,000 to 1,200	15 and 60	10 to 30	225 (1,200°C/60 min/15 MPa)	—
Ti-45Al/Ti-48Al (Yan and Wallach 1993)	900 to 1,200	16 to 50	20 to 150	—	260 (1,200°C/16 min/20 MPa)
Ti-48Al-2Mn-2Nb (Godfrey et al. 1993)	1,200 to 1,350	15 and 45	10	—	—
Ti-47Al-2Cr-0.2Si (Glatz and Clemens 1997)	1,000	60 to 180	20 to 40	608 (1,000°C/180 min/40 MPa)	—
Ti-47Al-4.5 (Cr,Mn,Nb,Si,B) (Çam and Koçak 1999)	950 to 1,100	60 to 180	20 to 40	—	373 (1,100°C/180 min/20 MPa) 466 (1,000°C/180 min/40 MPa PBHT (1,430°C/30 min))
Ti-47Al-3.7(Nb, Cr, C) (Çam et al. 1999)	950 to 1,000	60 to 3,000	5 to 40	—	542 (1,000°C/300 min/20 MPa)
Ti-47Al-3.7(Nb, Cr, C) (Çam et al. 2006)	950 to 1,000	60 to 300	5 to 20	—	383 (1,000°C/180 min/20 MPa) 580 (1,000°C/180 min/20 MPa PBHT (1,430°C/30 min))
Ti-45Al-5Nb-0.2C-0.2B (Holmquist et al. 1998)	1,000	15 to 120	20 to 100	850 (1,000°C/120 min/20 MPa)	—
Ti-45Al-7Nb-0.3W (Zhao et al. 2016)	1,000 to 1,200	15 to 60	15 to 30	657 (1,200°C/60 min/15 MPa)	—

PBHT: Post-Bond Heat Treatment.

through the interface. During creep rupture tests, it became evident that the bonding line had a strong effect on the nucleation and propagation of creep fracture.

Çam et al. (2008) investigated the diffusion bonding between Ti-47Al-3.7(Nb, Cr, C) at.% and Ti6Al4V alloys. The processing temperatures were 825, 850 and 875°C with bonding times between 15 and 45 min and a constant pressure of 5 MPa. Sound joints were obtained under various processing conditions except for bonds processed under less demanding conditions (825°C during 15 min), which showed isolated pores along the joint interface. The presence, at the interface, of a reaction layer composed of α_2-Ti_3Al grains was detected for all bonds, which led to the embrittlement of the joints. The thickness of this layer increased with increase in the bonding time. Some enrichment in niobium (from the γ-TiAl alloy) and vanadium (from Ti6Al4V) in the reaction layer was also observed. The mechanical properties of the joints were evaluated using shear and hardness tests. The joint processed at 850°C at a pressure of 5 MPa showed the highest value of shear strength (483 MPa); this value was only about 30 per cent of the strength of both base materials. The specimens fractured through the interface, which is consistent with the lower values of shear strength. This behaviour was associated with the presence of the brittle phase confirmed by the decrease in shear strength with the increase in bonding time.

Wang et al. (2006) have studied the possibility of joining Ti-46.5Al-2.5V-2Cr-1.5Nb at.% to a titanium alloy (Ti6Al4V) at 800, 880 and 900°C at a pressure of 100 MPa for 120 min. For all processing temperatures, interfaces free from defects were formed. For higher temperatures grain growth of the base material was observed. The mechanical properties were evaluated using three-point bending tests. The joints bonded at 880 and 900°C for 120 min showed high bending strength with the fracture located at the γ-TiAl alloy.

Wang et al. (2013) produced Ti-43Al-9V/Ti6Al4V joints by diffusion bonding. The joints were processed at 920°C for 120 min under a pressure of 45 MPa. Sound joints were apparently achieved. The thickness of the interfacial zone was about 11.2 μm and the interface showed a complex phase sequence with four zones: Ti-43Al-9V(base material)/γ(TiAl)/B2(ordered β(Ti))/α_2-Ti_3Al/α(Ti)/Ti-6Al-4V(base material).

Kanai et al. (2004, 2005) studied the diffusion bonding of γ-TiAl alloy (Ti-46Al-1.3Fe-1.2V-0.3B at.%) to three different titanium alloys: pure titanium, Ti6Al4V and Ti-17 (Ti-5.03Al-3.82Cr-1.90Sn-1.59Zr wt.%). The bonding between the γ-TiAl and Ti–17 alloys was performed at temperatures ranging from 700 to 1,075°C for 60 min with bonding pressures of 4.9 or 9.8 MPa. A tensile strength of 420 MPa, with the fracture occurring through the γ-TiAl base material, was obtained for joints processed at 1,000°C for 60 min with 9.8 MPa or at 1,075°C for 60 min with 4.9 MPa. The joints between the γ-TiAl alloy and pure titanium or Ti6Al4V consistently fractured at the interface.

The authors associated this different mechanical response of the joints with the microstructure of the interface. The formation of lamellar α_2-Ti_3Al layers in the γ-TiAl/Ti and γ-TiAl/Ti6Al4V joints was associated with the increased brittleness of these interfaces. This layer was suppressed in the γ-TiAl/Ti-17 interface, in which γ-TiAl grains and small particles of α_2-Ti_3Al were formed. These differences were explained by the composition of the Ti-17 alloy, in particular by the presence of chromium and molybdenum, which promoted the formation of γ-TiAl instead of α_2-Ti_3Al.

The studies described above demonstrated that solid-state diffusion bonding is a viable process for joining γ-TiAl alloys to themselves and to other materials. However, stringent conditions or complex manufacturing techniques are required for achieving mechanical properties similar to those of the base material. This makes the process less attractive economically. The use of interlayers is an option for making the diffusion bonding of γ-TiAl more attractive industrially. These interlayers can promote a reduction in processing conditions (lower-bonding temperature, dwell time and pressures) without impairing the mechanical properties.

3.3.1 Diffusion Bonding of γ-TiAl Alloys using Interlayers

As the name suggests, interlayers are layers that are placed between the mating surfaces of the materials to be joined. They can be introduced in various forms (foils, films or powders) with the intention of reducing the processing conditions, as stated above. These interlayers are usually pure soft metal layers and may have functions such as: to promote a localised diffusion, accommodate thermal stresses, and minimise problems with the roughness and quality of the surfaces. Some studies have reported the use of interlayers in the diffusion bonding of γ-TiAl alloys (Yan and Wallach 1993, Cao et al. 2007).

Yan and Wallach (1993) investigated the use of various interlayers (titanium, chromium, vanadium, manganese, niobium and molybdenum) deposited onto base material by sputtering and with thicknesses ranging from 0.5 to 1.5 μm in the diffusion bonding of a Ti-48Al at.% alloy. The use of interlayers improved the bonding of this γ-TiAl alloy—allowed it to be performed at comparatively lower temperatures and pressures. The interdiffusion between the interlayers and the γ-TiAl alloy was only partial. This originated a heterogeneous microstructure at the interface, different from that of the base material, thus compromising the mechanical response of the joint.

Cao et al. (2007) studied the diffusion bonding of γ-TiAl using a mixture of high purity titanium, aluminum, and carbon powders. The powders were cold pressed into disc specimens 0.5 mm thick. The discs were placed between the γ-TiAl alloys and heated up to 660ºC under pressures varying from 15 to 55 MPa. This compacted powder mixture reacted and formed a $TiAl_3$ layer at the interface, and γ-TiAl and TiC compounds at the central region. However, lower strength values were achieved due to the porosity existing at the interface (not less than 6 per cent vol.) and the formation of brittle phases.

In another study (Cao et al. 2006), the authors used an Ag-based brazing alloy to assist the bonding and increase the joint quality. The brazing foil, with a thickness of 100 μm, was inserted between the compacted powders and the γ-TiAl alloy. With the application of the brazing alloy, porosity decreased and joining strength increased. However, the presence of Ag-rich phases made this joint impractical for high temperature applications.

Titanium and/or nickel-thin foils were used as interlayers for diffusion bonding of Ti-45Al-5Nb at.% alloy.

Figure 3.4 shows the microstructure of the joints processed with a titanium foil (5 μm thick and 99.6 per cent purity) at 800 and 900°C for 60 min under a pressure of 5 MPa.

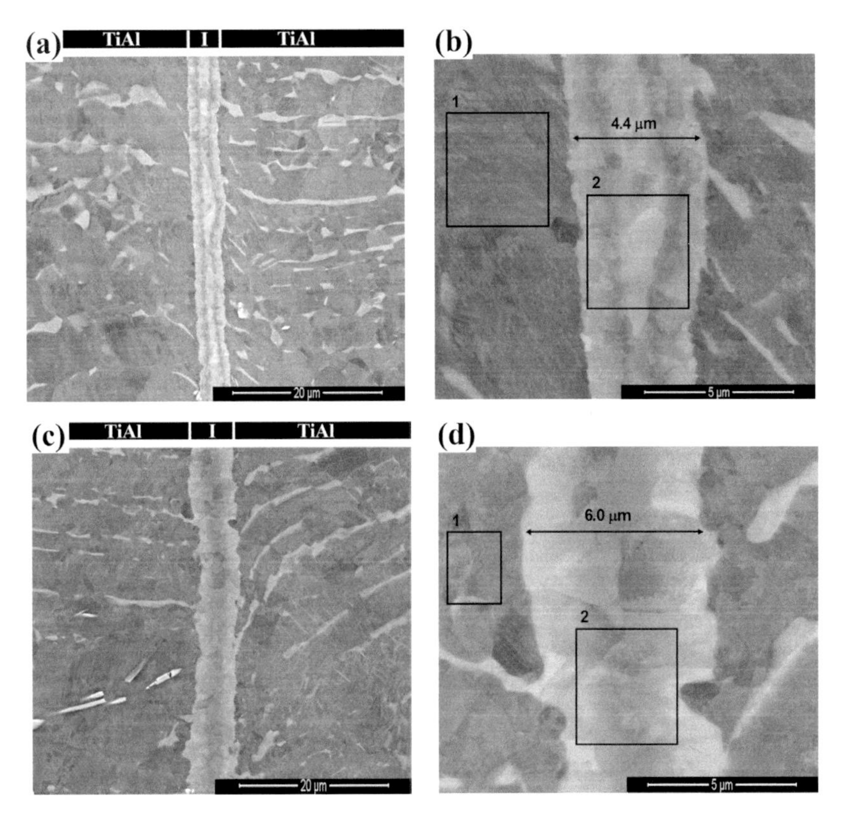

Figure 3.4 Scanning Electron Microscopy (SEM) images of diffusion bonds of γ-TiAl with a titanium foil (5 μm thick and 99.6 per cent purity) as interlayer processed during 60 min under a pressure of 5 MPa at: (a) and (b) 800°C and (c) and (d) 900°C. In (b) and (d) the areas analysed by Energy-Dispersive X-Ray Spectroscopy (EDS) are indicated.

At 800ºC, unbonded areas were detected, but with the increase in temperature these areas disappeared. However, some pores were observed between the base material and the titanium foil even at the joints produced at 900ºC. In the joints processed at 800°C, two distinct layers were observed at the interface, whereas in those processed at 900°C only one layer composed of equiaxed grains was detected. The total thickness of the interface increased from 4.4 to 6.0 μm with an increase in temperature from 800 to 900°C. The chemical composition of the layers was determined by Energy-Dispersive X-Ray Spectroscopy (EDS) analysis and the results are presented in Table 3.3, the analysed zones 1 (base material) and 2 (interface) being marked in Figure 3.4.

These results showed an intense diffusion of niobium and aluminum through the interface, these elements reacted with the titanium foil to produce the brittle α_2-Ti_3Al phase. The diffusion increased in line with increasing temperature; aluminum and niobium content in the center of the interface increased from 16.6 to 26.8 per cent and from 1.6 to 2.0 per cent, respectively, when the temperature increased from 800 to 900°C.

Joints processed using a nickel foil (1 μm thick and 99.95 per cent purity) with the same processing conditions were unsuccessful. The fracture of the joints occurred during the metallographic preparation of the bond cross-sections, revealing a reduced mechanical strength.

Diffusion bonding of γ-TiAl alloys assisted by an interlayer consisting of a nickel foil (1 μm thick and 99.95 per cent purity) surrounded by two titanium foils (5 μm thick and 99.6 per cent purity) was investigated at 900°C for 60 min under a pressure of 5 MPa (Figure 3.5). A bonded region of the interface is presented in Figure 3.5 ((a) and (b)). In the same figure, the zones analysed by EDS are indicated and their chemical composition is shown in Table 3.4.

The results indicate that the bonded regions resulted from reactions between the titanium and nickel foils, leading to the formation of a central layer (Zones 3 and 4) with a chemical composition close to that of the Ti_2Ni phase, and reactions of the titanium foil with the γ-TiAl alloy (a darker layer, Region 2) with a composition close to that of the α_2-Ti_3Al phase. Some nickel from the central foil diffused to the interface adjacent to the base materials, forming small Ni-rich areas with 8.5 per cent nickel (see Zone 1 in Figure 3.5).

Table 3.3 EDS analysis (at.%) of the areas identified in Figure 3.4.

Zones	800°C			Phases	900°C			Phases
	Al	Ti	Nb		Al	Ti	Nb	
1	47.7	47.3	5.0	γ-TiAl	45.7	49.1	5.3	γ-TiAl
2	16.6	82.0	1.4	α(Ti)+ α_2-Ti_3Al	26.8	71.2	2.0	α_2-Ti_3Al

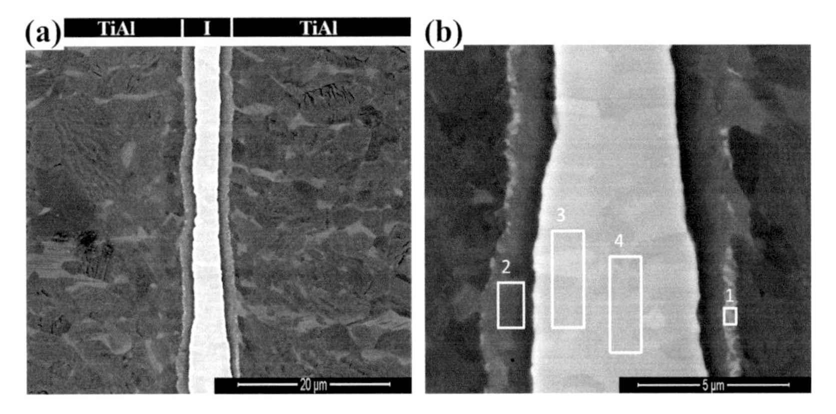

Figure 3.5 SEM images of diffusion bond of γ-TiAl with interlayers consisting of one nickel foil placed between two titanium foils. Joint processed at 800°C for 60 min under a pressure of 5 MPa.

Table 3.4 EDS analysis (at.%) of the areas identified in Figure 3.5.

Chemical composition (at.%)					Phases
Zones	Al	Ti	Ni	Nb	
1	37.4	49.4	8.5	4.7	α_2-Ti_3Al
2	28.5	66.7	1.4	3.4	α_2-Ti_3Al
3	2.7	63.9	33.4	—	Ti_2Ni
4	2.4	64.0	33.6	—	Ti_2Ni

Note that this layer, identified as the α_2-Ti_3Al phase, had a thickness less than the EDS volume of interaction, and therefore the results are only indicative, since they are affected by the composition of the surrounding areas. These joints showed unbonded zones, suggesting that diffusion between the foils of titanium and nickel and the γ-TiAl alloy was not sufficient to ensure bonding over the entire interface.

The use of thin foils as interlayers did not result in improvements in diffusion bonding of γ-TiAl. When a nickel foil assisted diffusion bonding, the bonds were intermittent, with a very low mechanical strength, which led to fracture during the metallographic preparation. The joining of γ-TiAl with a titanium foil or with a nickel foil between two titanium foils produced samples with bonded regions. However, defects such as pores, cracks and unbonded areas compromised the mechanical properties of these joints. Consequently, these foils of ductile metals are not suitable for successfully performing the solid-state diffusion bonding of γ-TiAl alloys at lower temperatures and pressures.

The use of interlayers was also investigated in the dissimilar diffusion bonding of γ-TiAl alloys. He et al. used foils of vanadium and copper (He et al. 2002) or titanium/vanadium/copper (He et al. 2003) to bond γ-TiAl to steel. The titanium and copper foils had thicknesses of 30 μm and the vanadium foil of 100 μm. Joints processed without the titanium foil fractured through the interlayer at relatively low tensile stress (maximum strength of 210 MPa).

Nickel interlayers were also used to assist the diffusion bonding of γ-TiAl to Ti_3SiC_2 ceramic (Liu et al. 2014). The nickel foil, with a thickness of 80 μm, had an important effect: it more than doubled the shear strength of the joints. Without the nickel foil, a maximum shear strength of 19.7 MPa was obtained for the joints processed at 1,000ºC for 60 min and at a bonding pressure of 30 MPa; the nickel foil increased the shear strength to 52.3 MPa, when bonding was performed at 950ºC for 60 min under 30 MPa. The improvement in the joint quality was associated with the modification of the phases formed at the interface, in particular with the disappearance of the very brittle Ti_5Si_3 phase.

A few studies have been undertaken using metallic interlayers in the diffusion bonding of γ-TiAl alloys. In general, these studies have proved that, the use of such interlayers introduces some improvements in solid-state diffusion bonding, although further work is required to reduce the demanding processing conditions that are industrially unattractive.

3.4 Concluding Remarks

Diffusion bonding is a solid-state process, which enables the successful joining of γ-TiAl alloys to themselves and to other materials. The joining of γ-TiAl alloys by this process is possible without the melting of the base materials, thus minimising the problems of hot cracking and high residual stresses. The process produces a very thin interface with uniform microstructure (a multilayered interface, as produced in brazing, is avoided) and the mechanical properties of the joint may be similar to the base material with proper selection of bonding parameters. However, the processing conditions are very demanding for industrial implementation, since high temperatures and pressures over long dwell time are required. The use of interlayers has proved to be one option for reducing the bonding conditions, but further studies are required. Despite the best efforts of the research community to obtain and develop processes for the joining of γ-TiAl alloys, new approaches are essential to establish a bond process that consistently produces sound joints, does not significantly affect the microstructure and properties of the base materials and ensures good mechanical properties of the joints, either at room temperature or at the service temperature.

Keywords: Aerospace; automotive; bonding temperature; diffusion; diffusion bonding; dissimilar joints; interdiffusion; interlayers; mating surfaces; similar joints; steel; superalloys; superplastic forming; Ti6Al4V.

3.5 References

Çam, G. and M. Koçak. 1999. Diffusion bonding of investment cast γ-TiAl. J. Mater. Sci. 34: 3345–3354.

Çam, G., H. Clemens, R. Gerling and M. Koçak. 1999. Diffusion bonding of γ-TiAl sheets. Intermetallics 7: 1025–1031.

Çam, G., G. Ipekoğlu, K.-H. Bohm and M. Koçak. 2006. Investigation into the microstructure and mechanical properties of diffusion bonded TiAl alloys. J. Mater. Sci. 41: 5273–5282.

Çam, G., U. Özdemir, V. Ventzke and M. Koçak. 2008. Microstructural and mechanical characterization of diffusion bonded hybrid joints. J. Mater. Sci. 43: 3491–3499.

Cao, J., J.C. Feng and Z.R. Li. 2006. Joining of TiAl intermetallic by self-propagating hightemperature synthesis. J. Mater. Sci. 41: 4720–4724.

Cao, J., J.C. Feng and Z.R. Li. 2007. Effect of reaction heat on reactive joining of TiAl intermetallics using Ti-Al-C interlayers. Scr. Mater. 57: 421–424.

Divinski, S. and Chr. Herzig. 2000. On the six-jump cycle mechanism of self-diffusion in NiAl. Intermetallics 8: 1357–1368.

Glatz, W. and H. Clemens. 1997. Diffusion bonding of intermetallic Ti-47Al-2Cr-0.2Si sheet material and mechanical properties of joints at room temperature and elevated temperatures. Intermetallics 5: 415–423.

Godfrey, S.P., P.L. Threadgill and M. Strangwood. 1993. High temperature phase transformation kinetics and their effects on diffusion bonding of Ti-48Al-2Mn-2Nb. J. Phys. IV 3: C7-485–C7-488.

He, P., J.-C. Feng, Y.Y. Qian and B.-G. Zhang. 2002. Microstructure and strength of TiAl/40Cr joint diffusion bonded with vanadium-copper filler metal. Trans. Nonferrous Met. Soc. China 12: 811–813.

He, P., J.C. Feng, B.G. Zhang and Y.Y. Qian. 2003. A new technology for diffusion bonding intermetallic TiAl to steel with composite barrier layers. Mater. Charact. 50: 87–92.

Hefti, L.D. 2008. Innovations in the superplastic forming and diffusion bonded process. J. Mater. Eng. Perform. 17: 178–182.

Herrmann, D. and F. Appel. 2009. Diffusion bonding of γ(TiAl) alloys: Influence of composition, microstructure, and mechanical properties. Metall. Mater. Trans. A-Phys. Metall. Mater. Sci. 40A: 1881–1902.

Herzig, Chr., T. Przeorski and Y. Mishin. 1999. Self-diffusion in γ-TiAl: an experimental study and atomistic calculations. Intermetallics 7: 389–404.

Holmquist, M., V. Recina, J. Ockborn, B. Pettersson and E. Zumalde. 1998. Hot isostatic diffusion bonding of titanium alloy Ti-6Al-4V to gamma titanium aluminide IHI Alloy 01A. Scr. Mater. 39: 1101–1106.

Holmquist, M., V. Recina and B. Pettersson. 1999. Tensile and creep properties of diffusion bonded titanium alloy IMI 834 to gamma titanium aluminide IHI alloy 01A. Acta Mater. 47: 1791–1799.

Imayev, R.M., V.M. Imayev and G.A. Salishchev. 1992. Formation of submicrocrystalline structure in TiAl intermetallic compound. J. Mater. Sci. 27: 4465–4471.

Kanai, S., S. Seto and T. Wada. 2004. Diffusion bonding of TiAl intermetallic compound to Ti-6Al-4V alloy and pure titanium. Q. J. Jpn. Weld. Soc. 22: 580–586.

Kanai, S., S. Seto and H. Sugiura. 2005. Tensile properties of diffusion bonds between TiAl intermetallic compound and titanium alloy. Mater. Trans. 46: 2484–2489.

Kestler, F. and H. Clemens. 2003. Production, processing and application of γ(TiAl)-based alloys. pp. 351–388. *In*: C. Leyens and M. Peters [eds.]. Titanium and Titanium Alloys:

Fundamentals and Applications. Wiley-VCH Verlag GmbH & Co. KGaA, Weinheim, Germany.
Liu, J., J. Cao, X. Song and J. Feng. 2014. Investigation on diffusion bonding of TiAl intermetallic to Ti_3AlC_2 ceramic with Ni interlayer. China Weld. 23: 75–78.
Lutfullin, R. Ya., R.M. Imayev, O.A. Kaibyshev, F.N. Hismatullin and V.M. Imayev. 1995. Superplasticity and solid state bonding of TiAl intermetallic compound with micro- and submicrocrystalline structure. Scr. Metall. Mater. 33: 1445–1449.
Lütjering, G. and J.C. Williams. 2007. Titanium. 2nd edition. Springer-Verlag Berlin, Heidelberg, Germany.
Mehrer, H. 2007. Diffusion in Solid: Fundamentals, Methods, Materials, Diffusion-Controlled Processes, Springer-Verlag Berlin, Heidelberg, Germany.
Messler Jr., R.W. 2004. Principles of Welding: Processes, Physics, Chemistry, and Metallurgy. Wiley-VCH Verlag GmbH & Co. KGaA, Weinheim, Germany.
Nakajima, H., K. Nonaka, W. Sprengel and M. Koiwa. 1997. Self-diffusion and interdiffusion in intermetallic compounds. Mater. Sci. Eng. A-Struct. Mater. Prop. Microstruct. Process. 239–240: 819–827.
Nakao, Y., K. Shinozaki and M. Hamada. 1991. Diffusion bonding of intermetallic compound TiAl. ISIJ Int. 31: 1260–1266.
Wang, X.-F., M. Ma, X.-B. Liu, X.-Q. Wu, C.-G. Tan, R.-K. Shi et al. 2006. Diffusion bonding of γTiAl alloy to Ti-6Al-4V alloy under hot pressure. Trans. Nonferrous Met. Soc. China 16: 1059–1063.
Wang, X.R., Y.Q. Yang, X. Luo, W. Zhang, G.M. Zhao and B. Huang. 2013. An investigation of Ti-43Al-9V/Ti-6Al-4V interface by diffusion bonding. Intermetallics 36: 127–132.
Wu, G.Q. and Z. Huang. 2001. Superplastic forming/diffusion bonding of laser surface melted TiAl intermetallic alloy. Scr. Mater. 45: 895–899.
Wu, G.Q. and Z. Huang. 2003. Weldability and microstructure of laser-surface-remelted TiAl intermetallic alloy. Mater. Sci. Eng. A-Struct. Mater. Prop. Microstruct. Process. 345: 286–292.
Wu, G.Q., Z. Huang, C.Q. Chen, Z.J. Ruan and Y. Zhang. 2004. Superplastic diffusion bonding of γ-TiAl-based alloy. Mater. Sci. Eng. A-Struct. Mater. Prop. Microstruct. Process. 380: 402–407.
Xun, Y.W. and M.J. Tan. 2000. Applications of superplastic forming and diffusion bonding to hollow engine blades. J. Mater. Process. Technol. 99: 80–85.
Yan, P. and E.R. Wallach. 1993. Diffusion-bonding of TiAl. Intermetallics 1: 83–97.
Zhao, K., Y. Liu, L. Huang, B. Liu and Y. He. 2016. Diffusion bonding of Ti-45Al-7Nb-0.3W alloy by spark plasma sintering. J. Mater. Process. Technol. 230: 272–279.
Zhou, Y. [ed.]. 2008. Microjoining and Nanojoining. Woodhead publishing and Maney Publishing on behalf of The Institute of Materials, Minerals & Mining, CRC Press, Woodhead Publishing Limited, Cambridge, England, UK.
Zhu, H., B. Zhao, Z. Li and K. Maruyama. 2005. Superplasticity and superplastic diffusion bonding of a fine-grained TiAl alloy. Mater. Trans. 46: 2150–2155.

CHAPTER 4

Reactive Multilayers as Interlayer for Joining Processes

4.1 Introduction

Reactive multilayers are usually composed of two different materials that react exothermically with each other at well-defined temperatures. The layers have nanometric or submicrometric thicknesses and the reaction locally releases large amounts of stored chemical energy (for a recent review on this topic see Adams 2015).

The nanometric character and its highly energetic reaction suggest that its use as an interlayer is advantageous in joining processes. Diffusion bonding and brazing are two joining processes that can be improved by the use of an interlayer of reactive multilayers. In brazing, these can be used to melt the brazing alloy at lower processing temperatures (Swiston Jr. et al. 2003, 2005, Trenkle et al. 2008, Boettge et al. 2010). In diffusion bonding, reactive multilayers can simultaneously function as a localised heat source and increase diffusivity due to their nanometric character. Their use in the joining of hard-to-join materials has been proven to be effective (Duarte et al. 2006, Ustinov et al. 2008, Simões et al. 2010b, 2013).

The selection of the multilayer system to be used in diffusion bonding in order to produce a successful bond depends on the heat release and enhanced diffusion, reaction between the elements of the base materials and the multilayers. Particularly important are the phases formed during the reaction. The strategy for the selection of this interlayer is to use a chemical composition similar to that of base materials or to use elements which prevent the formation of brittle phases that will affect the mechanical properties. In any case, there are many systems that can benefit the diffusion bonding process, and so it is important to characterise the solid-state reactions occurring when using a reactive multilayer.

This chapter will consist of a description of the most promising reactive multilayers to be used as an intermediate layer in joining processes, in particular for the joining of γ-TiAl alloys. The microstructural features of these different multilayers, as well as their anisothermal solid-state reactions, will be explored.

4.2 Reactive Multilayers

Reactive multilayers consists of layered materials that react at well-defined temperatures (Ramos et al. 2006, 2009, 2014, Adams 2015). Although a variety of reactive multilayers have been reported (Michaelsen et al. 1994, 1995, 1996, Ramos et al. 2006, 2009, 2014, Weihs 2014, Adams 2015, Woll et al. 2016) such as those of the metal-semiconductor, metal-oxide, metal-metalloid, metal-organic and metal-metal, this chapter will only analyse the last type of multilayers.

The atoms of the alternated layers mix and react by means of thermal explosion, thermal annealing or ignition of a self-propagating reaction. The thermal explosion occurs when the increase in temperature of the multilayer is above a critical heating rate, which promotes the simultaneous reaction of the entire multilayer. The reaction may also be initiated by local application of an energy pulse, which may be a laser or an electric discharge. In this case, the most typical one for reactive multilayers, a self-propagating reaction occurs in a short period of time, with a high heat release rate. The reactions are self-propagating if the atomic diffusion and energy release are sufficiently rapid. These reactive multilayers can be used as a local heat source, since a small energy pulse can release a large amount of heat and temperatures exceeding 2,000°C can be reached in adiabatic conditions (Woll et al. 2016).

The reaction by thermal annealing occurs when the multilayers are heated at a low heating rate in a furnace. In this case, depending on chemical composition and bilayer thickness, the reaction can occur in steps, forming intermediate phases prior to the final products (Lucadamo et al. 1999, 2001, Ramos and Vieira 2005, 2012, Ramos et al. 2006, 2009, 2014, Chen et al. 2007, Yajid et al. 2008, Weihs 2014, Adams 2015, Murphy et al. 2015, Woll et al. 2016). Multilayers that transform directly into the final products are those exhibiting a higher and quicker heat release, and which are therefore the most effective in improving the joining processes.

Exothermic reactions have been reported for various multilayer systems. Among the systems that exhibit higher heat release with higher propagation velocity of the reaction are those that react to form aluminides, such as Nb/Al (Lucadamo et al. 1999, 2001), Cu/Al (Yajid et al. 2008), Pt/Al (Murphy et al. 2015), Pd/Al (Ramos and Vieira 2012), Ru/Al (Woll et al. 2016), Ti/Al (Michaelsen et al. 1994, 1995, Banerjee et al. 1999, Rogachev et al. 2004, Ramos and Vieira 2005, Gachon et al. 2005, Ramos et al. 2006, 2009, Rogachev 2008, Illeková et al. 2008) and Ni/Al (Ma et al. 1989, 1990, 1991, Edelstein et al. 1994, 1995, Colgan et al. 1995, Michaelsen et al. 1996, Barmak et al. 1997, da Silva Bassani et al. 1997, Gavens et al. 2000, Jeske et al. 2003, Blobaum et al. 2003, Lee et al. 2005, Trenkle et al. 2005, 2010, Qiu and Wang 2007, Noro et al. 2008, Morris et al. 2010, Simões et al. 2010a, 2011, Politano et al. 2013, Rogachev et al. 2014, 2016).

Systems that form titanides, in particular Fe/Ti (Chen et al. 2007) and Ni/Ti (Clemens 1986, Bhatt et al. 2006, Cavaleiro et al. 2014, 2015), have also been the object of several studies. Some of these systems have proved their efficiency in improving bonding by reducing the bonding processing conditions (temperature, dwell time and pressure). Special attention has been given to Ni/Al (Swiston Jr. et al. 2003, 2005, Trenkle et al. 2008, Simões et al. 2010b, 2013, 2014, 2016a,d), Ti/Al (Duarte et al. 2006, Ustinov et al. 2008, Cao et al. 2012) and Ni/Ti systems (Simões et al. 2016b,c, Emadinia et al. 2016).

The successful application of these multilayers is associated with the heat released during the reaction and the chemical composition, as mentioned above. In turn, the heat released is strongly dependent on the intermixing/interdiffusion that occurs during the production of the multilayers. A smaller intermixing leads to increased heat release when the multilayer reacts. The production technique and the bilayer thickness have a marked effect on intermixing.

According to the aforementioned studies, the reaction in multilayer metal/aluminum often occurs with the formation of intermediate phases (Clemens 1986, Ma et al. 1989, 1990, 1991, Edelstein et al. 1994, 1995, Colgan et al. 1995, Michaelsen et al. 1995, 1996, Swiston Jr. et al. 2003, 2005, Trenkle et al. 2005, 2010, Bhatt et al. 2006, Qiu and Wang 2007, Illeková et al. 2008, Noro et al. 2008, Morris et al. 2010, Simões et al. 2011, 2013, Politano et al. 2013, Cavaleiro et al. 2014, 2015, Ramos et al. 2014, Rogachev et al. 2014, 2016). The reaction path depends on processing technology, chemical composition, bilayer thickness and on the conditions of reaction activation (in particular, the heating rates), as will be analysed in detail in the following subchapters.

The reactive multilayers can be produced by various techniques: Physical Vapor Deposition (PVD) (Clemens 1986, Ma et al. 1989, 1990, 1991, Edelstein et al. 1994, 1995, Colgan et al. 1995, Michaelsen et al. 1995, 1996, Barmak et al. 1997, da Silva Bassani et al. 1997, Swiston Jr. et al. 2003, 2005, Trenkle et al. 2005, 2010, Bhatt et al. 2006, Qiu and Wang 2007, Illeková et al. 2008, Noro et al. 2008, Morris et al. 2010, Simões et al. 2011, 2013, Politano et al. 2013, Cavaleiro et al. 2014, 2015, Ramos et al. 2014, Rogachev et al. 2014, 2016) and the cold rolling process (Sieber et al. 2001, Ding et al. 2005, Jamaati and Toroghinejad 2011, Sun et al. 2011, Ghalandari et al. 2016) are two representative examples.

However, most studies indicate that PVD processes are the most promising for obtaining multilayers with uniform nanometric layers. The production of alternating layers by deformation processes is limited to a few systems, since it depends on many requirements, especially the mechanical behaviour of the materials.

Sputter deposition is a PVD method widely used in the production of multilayers, since it is a versatile technique allowing easy optimisation of deposition parameters. The reactive multilayers can be produced by sputtering, using two targets with the composition of the materials of the layers. For applications in joining processes, reactive multilayers can be deposited on the base materials or may be used as a freestanding foil. The surface preparation of the base materials is very important when they are to be coated with multilayers. This preparation affects the adhesion between the base material and multilayers, which is essential for subsequent joining processes. The sputtered multilayer films have a very good adhesion to the base material and replicates the surface roughness and any surface defect. During the deposition process, the processing conditions that need to be considered are the atmosphere used, the polarisation of the substrates, the power densities of the targets, the substrate holder rotation speed, the time of the deposition, the distance of the holder to the targets and the use of a holder able to dissipate heat. The rotation speed of the substrate holder and power densities of the targets are selected and optimised to produce multilayers with the desired layer thicknesses and chemical composition.

An important characteristic of the reactive multilayers is their bilayer thickness (Λ), which is the sum of the thickness of one layer of each material and is commonly referred to as the modulation period. The deposition time determines the final thickness of the sputtered multilayer film.

In Figure 4.1 a schematic illustration and a Scanning Electron Microscopy (SEM) image of a reactive multilayer produced by sputtering are presented.

Accumulative Roll Bonding (ARB) is a severe plastic deformation process that was also reported for the production of multilayers. The ARB process is a simple and an inexpensive option for such production. The main advantage of this process is the production of thick multilayers with nanometric or ultrafine layers.

For producing multilayers by this process, foils of pure materials are stacked and rolled. During rolling, the foils of the two materials are bonded to each other. After the first rolling, the foil is cut in two, the halves overlapped and subjected to another rolling cycle. The passes are repeated until a layered structure, having the desired layer thickness, is produced. The main challenges of producing nanometric multilayers by this technique is to ensure a good adhesion between the layers and to be able to perform a great number of rolling passes without the fracture of the foil. However, for some systems it is very difficult to obtain nanometric and continuous layers due to the large difference in the mechanical behaviour of the two materials.

An example of these systems is Ni/Al reactive multilayers, for which it was not possible to produce continuous multilayers with reduced thicknesses. In fact, the increase in ARB cycles promoted the formation of nickel islands in an aluminum matrix instead of alternated layers, see Figure 4.2. During ARB, the harder material (nickel) necked and broke into fragments (Sieber et al. 2001, Simões et al. 2016a). Ni/Ti and Ni/Al multilayers have been applied to joining some materials (Emadinia et al. 2016, Simões et al. 2016a) and exhibited some advantages over the commercial interlayer.

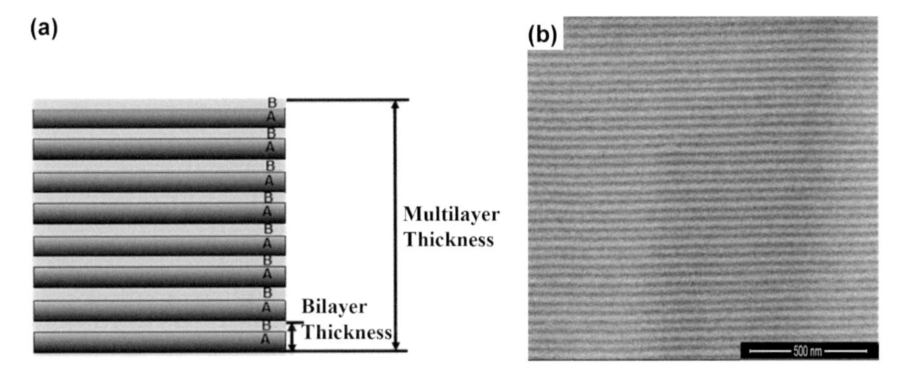

Figure 4.1 (a) Schematic illustration of a reactive multilayer composed of two different materials and (b) SEM image of a reactive multilayer produced by sputtering.

Figure 4.2 SEM image of an Ni/Al reactive multilayer produced by ARB.

As previously mentioned, the successful use of reactive multilayers as a bonding aid depends on the capacity to deliver heat directly at the bonding interface and on the amount of heat released by the exothermic reaction of the multilayers. The most reported reactive multilayers used to improve bonding of γ-TiAl alloys will be described in detail: Ti/Al, Ni/Al and Ni/Ti. Characteristics of these multilayers, such as bilayer thickness, chemical composition, multilayer thickness and prior atomic intermixing/interdiffusion between the layers control the exothermic reaction and will be discussed below.

4.2.1 Ti/Al Reactive Multilayers

The Ti/Al reactive multilayers were the first to be studied for application to the diffusion bonding of γ-TiAl alloys (Duarte et al. 2006, Ustinov et al. 2008). The use of these multilayers seems to be a good strategy for improving the diffusion at the bonding interface without causing discontinuities in chemical composition.

Ti/Al reactive multilayers can be produced by magnetron sputtering with a bilayer thickness from 4 nm to several micrometers (Michaelsen et al. 1994, 1995, Banerjee et al. 1999, Rogachev et al. 2004, Gachon et al. 2005, Ramos and Vieira 2005, Ramos et al. 2006, 2009, Illeková et al. 2008, Rogachev 2008). These multilayers are characterised by a columnar morphology that becomes more pronounced as the bilayer thickness decreases. Microstructural characterisation of the multilayers revealed a well-defined layered microstructure composed of grains that are no larger than the layers' thickness. Therefore, nanometric grains constitute the multilayers produced with a nanometric bilayer thickness. The nanometric grain size is identified as responsible for the increased diffusivity in these multilayers, the diffusion through the grain boundaries being the preferential diffusion mechanism (Rogachev et al. 2004, Illeková et al. 2008, Rogachev 2008).

Intermixing during deposition is not reported for these multilayers. The results of the characterisation of the as-deposited multilayer films by high-resolution techniques only detected the presence of titanium and aluminum without the formation of other phases (Ramos et al. 2014). The intermixing is touted as one of the parameters which can decrease the heat released during the exothermic reaction. As the Ti/Al multilayers undergo no intermixing, it is expected that the heat released by this system reaction will depend only on the bilayer thickness and the chemical composition.

Table 4.1 summarises the influence of chemical composition and bilayer thickness in the phase formation and temperature of the exothermic reaction reported in the literature. Another characteristic that affects the amount of heat released by the reaction is the formation of intermediate phases. When the multilayers react in several steps, with the sequential formation of intermediate phases, the heat is slowly released in the temperature region between these steps and the heating effect is smaller.

Table 4.1 Influence of chemical composition and bilayer thickness of Ti/Al reactive multilayers on the reactions temperature and phase formation.

Chemical composition of alloy (at.%)	Bilayer thickness (nm)	Reaction temperature (°C)	Phases
Ti-48.5Al (Ramos and Vieira 2005, Ramos et al. 2006)	4	600	γ-TiAl + α_2-Ti_3Al
	20	575	γ-TiAl + α_2-Ti_3Al
	200	480	α(Ti) + $TiAl_3$
		575	γ-TiAl + α_2-Ti_3Al
	500	620	γ-TiAl + $TiAl_3$
Ti-50Al (Illeková et al. 2008)	8	450	γ-TiAl + α_2-Ti_3Al
	40	400	γ-TiAl + α_2-Ti_3Al
	210	480	α(Ti) + $TiAl_3$
		580	γ-TiAl + α_2-Ti_3Al
Ti-50Al (Rogachev et al. 2004, Gachon et al. 2005)	110	320	α(Ti) + $TiAl_3$
		700	γ-TiAl + α_2-Ti_3Al
Ti-60Al (Rogachev et al. 2004, Gachon et al. 2005)	144	280	α(Ti) + $TiAl_3$
		380	γ-TiAl + α_2-Ti_3Al

For the exothermic reaction of the multilayers to lead to the formation of only the γ-TiAl phase it is essential that the as-deposited multilayers present an equiatomic chemical composition. The chemical composition of the multilayers is a crucial factor and will affect the phases formed and consequently the heat release during the reaction. Gachon et al. (2005) reported reaction products rich in $TiAl_3$ for Ti/Al multilayers.

For Ti/Al reactive multilayers with near equiatomic chemical composition, the number of exothermic reactions and phase formation are strongly related to bilayer thickness. For large bilayer thicknesses, exceeding 100 nm, two exothermic reactions were detected, while only one was observed for smaller bilayer thickness.

Ramos et al. (Ramos and Vieira 2005, Ramos et al. 2006) reported that for Ti-48.5Al at.% multilayers with a bilayer thickness larger than or equal to 200 nm, the Differential Scanning Calorimetry (DSC) curves exhibited two exothermic peaks. The first peak was associated with the formation of the Al-rich phase, $TiAl_3$. The second peak was associated with the formation of γ-TiAl and a small amount of α_2-Ti_3Al. For the multilayers with a small bilayer thicknesses, 4 and 20 nm, only one peak was observed in the DSC curve at 600°C, corresponding to the direct formation of γ-TiAl and small amounts of α_2-Ti_3Al. Despite the formation of α_2-Ti_3Al in the final stage, the major phase is γ-TiAl for the multilayers studied. The temperature of the exothermic reactions decreased with the increase in bilayer thickness, but remained below 700°C for all bilayer thicknesses.

Illeková et al. (2008) reported similar results for the phase formation of the Ti/Al reactive multilayers with an equiatomic chemical composition produced with 8 to 2,000 nm of bilayer thickness. γ-TiAl and α_2-Ti_3Al were detected for all multilayers and were independent of bilayer thickness. For the larger bilayer thicknesses, greater than 56 nm, two exothermic reactions occurred: $TiAl_3$ formed at low temperatures and γ-TiAl and small amounts of α_2-Ti_3Al at high temperatures. The heat of the reactions determined by these authors was high and decreased in proportion to the bilayer thickness, varying from -640 to -300 J/g (the sum of the heat of all the reactions) to -100 J/g for multilayers with a bilayer thickness of 8 nm. Although the multilayers with large bilayer thicknesses release a large amount of heat, in fact the reaction of formation of γ-TiAl in one step released a much lower heat than the expected value of -480 J/g (Adams 2015).

Michaelsen et al. (1994, 1995) demonstrated the formation of $TiAl_3$ for Ti/Al reactive multilayers composed of micrometric layers. A kinetic study revealed a slight decrease in the activation energy with the bilayer thickness. The phase evolution of Ti/Al multilayers under slow heating rates was also studied by Rogachev (2008) with similar results.

Drawing on these works, it is clear that the bilayer thickness and chemical composition of Ti/Al reactive multilayers have a strong influence on the phase formation during the exothermic reactions. Although the heat released by the reaction of these multilayers is not very high, their use as an interlayer in diffusion bonding is nonetheless interesting, since their composition can be designed to match the composition of the γ-TiAl alloys to be joined. The small bilayer thicknesses appear to be the most favourable, since the multilayer reaction occurs in only one step. However, it is important to consider that small amounts of the α_2-Ti_3Al phase were also formed, which can negatively influence the properties of the joints.

4.2.2 Ni/Al Reactive Multilayers

Ni/Al reactive multilayers are one of the most promising systems for use as an interlayer, due to the high reactivity and large amount of heat released by the reaction. The Ni/Al reactive multilayers began to be recognised for their application as an interlayer in the bonding of several materials, especially γ-TiAl alloys, because of the good results obtained in glass and ceramics bonding (Swiston Jr. et al. 2003, 2005). Although the chemical composition of these multilayers is quite different from that of the base material, the benefit of its use in diffusion bonding of γ-TiAl alloys results from the large amount of heat released during the exothermic reaction of the multilayer, much larger than the amount released by Ti/Al reactive multilayers. Furthermore, the exothermic reaction of these multilayers leads to the formation of intermetallic phases, NiAl or Ni_3Al, which exhibits interesting properties and do not restrict the high-temperature applications of the alloys.

The Ni/Al reactive multilayers can also be produced successfully by magnetron sputtering (PVD process) with bilayer thicknesses from 5 nm to several micrometers (as the Ti/Al reactive multilayers), and with a chemical composition adjusted to the application intended (Ma et al. 1989, 1990, 1991, Edelstein et al. 1994, 1995, Colgan et al. 1995, Michaelsen et al. 1996, Barmak et al. 1997, da Silva Bassani et al. 1997, Gavens et al. 2000, Blobaum et al. 2003, Jeske et al. 2003, Rogachev et al. 2004, 2014, 2016, Lee et al. 2005, Qiu and Wang 2007, Illeková et al. 2008, Noro et al. 2008, Rogachev 2008, Ramos et al. 2009, Morris et al. 2010, Simões et al. 2010a, 2011, Politano et al. 2013).

Figure 4.3 shows SEM images of Ni/Al reactive multilayer, with an equiatomic composition, produced by sputtering with a 30 nm bilayer thickness. The alternated nanometric layers of nickel (bright layers) and aluminum (dark layers) are clearly visible in these images. As observed for the Ti/Al reactive multilayers, the microstructure is characterised by a columnar morphology.

A more in-depth investigation of the Ni/Al reactive multilayers, using Transmission Electron Microscopy (TEM) and High Resolution TEM (HRTEM) analysis, revealed individual layers of nickel and aluminum consisting of nanometric grains (Ramos et al. 2009, Simões et al. 2011). Ramos et al. (2009) observed that the grain size of the nickel and aluminum layers is always smaller than the films' individual layer thickness.

Figure 4.4 shown the HRTEM and Fast Fourier Transform (FFT) of different areas of the Ni/Al reactive multilayer with a bilayer thickness of 5 nm. HRTEM images also revealed that aluminum and nickel nanometric grains comprise the respective layers though, in some localised areas at the

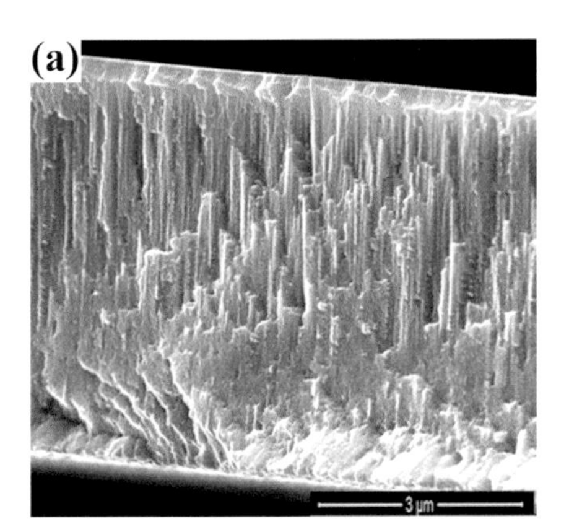

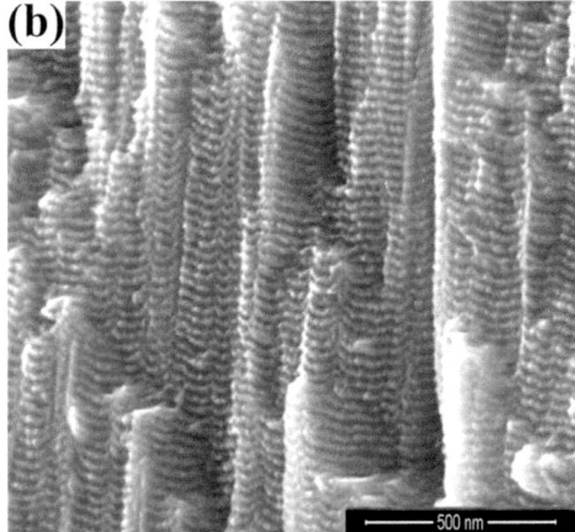

Figure 4.3 SEM images of an Ni/Al reactive multilayer produced by sputtering with a 30 nm bilayer thickness.

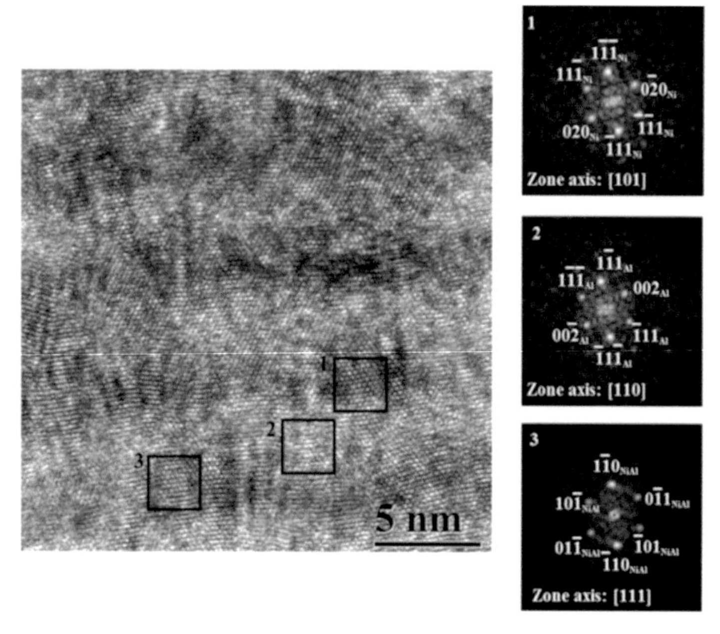

Figure 4.4 HRTEM image and FFT pattern of the three regions indicated. The dark layers are indexed as nickel (Region 1) and the bright layers as aluminum (Region 2), while in some localise regions is identified NiAl (Region 3).

interface between the layers, grains of the NiAl phase were also detected. Observation of the NiAl grains had shown the occurrence of intermixing and interdiffusion during the Ni/Al reactive multilayer deposition, which did not occur in the case of the Ti/Al reactive multilayers. In the Ni/Al multilayers a reaction between the layers occurred during the deposition. This intermixing was only detected for the multilayers with bilayer thicknesses smaller than 30 nm. Intermixing between nickel and aluminum layers and NiAl formation in as-deposited multilayers was observed in other studies; stable or metastable intermediate phases have been observed as a result of interdiffusion and phase transformations (Barmak et al. 1997, Gavens et al. 2000, Lee et al. 2005).

Intermixing was only observed in very localised and nanometric regions that can only be detected by high-resolution characterisation techniques; X-Ray Diffraction (XRD) was not sufficient for detecting intermixing and the formation of NiAl, as can be confirmed by observing the patterns of Figure 4.5, which only revealed peaks of nickel and aluminum in the multilayers of Figure 4.4.

The exothermic reaction between aluminum and nickel layers has been studied extensively. In the combustion mode, the nickel and aluminum layers transformed directly into NiAl (Rogachev et al. 2014, 2016), but in the

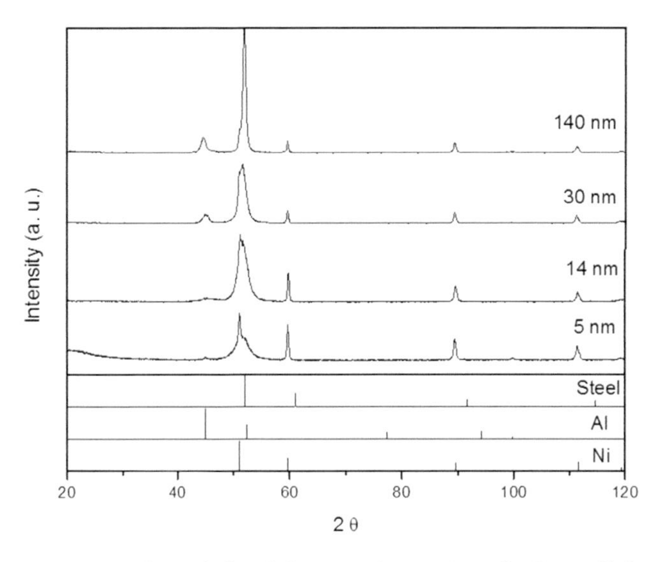

Figure 4.5 XRD patterns for Ni/Al multilayers with 5, 14, 30 and 140 nm of bilayer thickness deposited in a steel substrate.

early stages of the reaction $NiAl_3$ was formed at the reaction front. According to the literature, when the multilayers were transformed by annealing, the first products to form depends largely on the chemical composition of the multilayer being: NiAl (Edelstein et al. 1995, Michaelsen et al. 1996, da Silva Bassani et al. 1997, Noro et al. 2008), $NiAl_3$ (Edelstein et al. 1994, 1995, Michaelsen et al. 1996, da Silva Bassani et al. 1997, Sieber et al. 2001, Blobaum et al. 2003, Qiu and Wang 2007, Noro et al. 2008) or Ni_2Al_9 (Edelstein et al. 1994, 1995, da Silva Bassani et al. 1997, Barmak et al. 1997, Blobaum et al. 2003, Noro et al. 2008).

For all the cases, NiAl was the main product of the reaction for multilayers with an equiatomic chemical composition. In general, for small bilayer thicknesses the exothermic reaction occurred in only one step, forming the phase with a chemical composition close to the multilayer's average chemical composition (Edelstein et al. 1994, 1995); for larger bilayer thickness intermediate phase formation was observed and the reaction occurs in several steps. A summarisation of the results obtained for Ni/Al reactive multilayers produced by sputtering with different chemical compositions and bilayer thicknesses can be observed in Table 4.2.

The structural evolution of Ni/Al reactive multilayers with temperature has attracted the attention of some researchers. This evolution was investigated *in situ* by hot XRD (Michaelsen et al. 1996, Barmak et al. 1997, Noro et al. 2008) or by DSC combined with XRD and SEM observations (Michaelsen et al. 1996, da Silva Bassani et al. 1997, Nathani et al. 2007, Simões et al. 2011).

Table 4.2 Temperature of reactions and phase formation of Ni/Al reactive multilayers with different composition and bilayer thickness.

Chemical composition (at.%)	Bilayer thickness (nm)	Temperature of reactions (°C)	Phases
Ni-48 to 63Al (Michaelsen et al. 1996)	10	107	NiAl
	20	227	NiAl
		317	$NiAl_3$
Ni-47 to 52Al (Simões et al. 2011)	5	190	NiAl
	14	250	NiAl
	30	230	$NiAl_3$
		330	Ni_2Al_3 + NiAl
Ni-60Al (Barmak et al. 1997)	5	227	$NiAl_3$
	10	227	$NiAl_3$
	20	237	Ni_2Al_9
		300	$NiAl_3$
	320	277	Ni_2Al_9
		400	$NiAl_3$
Ni-60 to 64Al (Blobaum et al. 2003)	12.5	250	$NiAl_3$
	20	240	Ni_2Al_9
		300	$NiAl_3$
		320	Ni_2Al_3
	300	260	Ni_2Al_9
		330	$NiAl_3$
		370	Ni_2Al_3

As already mentioned in relation to Ti/Al reactive multilayers, the exothermic reactions of Ni/Al reactive multilayers are also strongly related to their chemical composition. Edelstein et al. (1994, 1995) characterised the exothermic reactions of $Ni_{0.6}/Al_{0.4}$ and $Ni_{0.25}/Al_{0.75}$ multilayers with bilayer thicknesses of 10, 80–100 and 400 nm. Their results revealed that the multilayers with 10 nm bilayer thickness reacted to form NiAl at 250°C and $NiAl_3$ at 285°C. For the larger bilayer thicknesses the DSC curves presented several peaks, Ni_2Al_9 formed at a lower temperature, while Ni_2Al_3 and $NiAl_3$ formed at higher temperatures for the 80 and 400 nm bilayer thicknesses, respectively.

The formation of these intermediate phases was also observed by Blobaum et al. (2003) in multilayers with an overall chemical composition designed to produce Ni_2Al_3 as the final product and a bilayer thickness larger than 12.5 nm. The first phase formed was Ni_2Al_9, following the growth of this phase, $NiAl_3$ and Ni_2Al_3 nucleated and grew sequentially. For the multilayer with a smaller bilayer thickness, as the intermixing was higher, the amount of aluminum was not sufficient to form Ni_2Al_9.

Michaelsen et al. (1996) investigated the reactions of Ni/Al reactive multilayers produced by magnetron sputtering with 10 and 20 nm bilayer thicknesses and with overall chemical composition ranging from 48 to 88 at.% of aluminum. DSC curves of all multilayers revealed a low-temperature reaction peak (around 170°C) of formation of NiAl. For the multilayers with chemical compositions of up to 63 at.% of aluminum and 10 nm of bilayer thickness, NiAl was the first phase formed by the reaction. The reactive multilayers with compositions higher than 63 at.% of aluminum revealed a sharp peak at around 300°C that corresponds to the formation of $NiAl_3$. For multilayers with 20 nm of bilayer thickness and chemical composition ranging from 48 to 72 at.% of aluminum, two reactions were identified, corresponding to the formation of NiAl and $NiAl_3$.

Barmak et al. (1997) also studied the reactions of the Ni/Al reactive multilayers produced by sputtering, with a chemical composition close to 75 at.% of aluminum and bilayer thicknesses ranging from 2.5 to 320 nm. For multilayers with a bilayer thickness smaller than 20 nm only one reaction was detected, from 227 to 247°C, corresponding to the formation of the $NiAl_3$ phase. Multilayers with larger bilayer thicknesses exhibited a second exothermic peak that became more dominant with increasing bilayer thickness, the first peak being associated with the interface-controlled growth and the second with the diffusion-controlled growth of the $NiAl_3$.

The formation of $NiAl_3$ in two stages was also reported by Qiu and Wang (2007). Another important observation of this work was the decrease in the heat of the reaction with the decrease in the bilayer thickness, interpreted as the result of intermixing and formation of NiAl grains during deposition, which was more important for multilayers with reduced bilayer thicknesses.

For the Ni/Al reactive multilayers with a chemical composition of 47–52 at.% of aluminum, the bilayer thickness is the factor that determines the amount of heat released during the exothermic reaction (Michaelsen et al. 1996, Barmak et al. 1997, da Silva Bassani et al. 1997, Simões et al. 2011). Multilayers with bilayer thicknesses smaller than 20 nm exhibited one exothermic reaction, the increase in the bilayer thickness promoting second or further reactions. For Ni/Al reactive multilayers with 5 and 14 nm bilayer thicknesses, the layers reacted at around 200 to 250°C to directly form NiAl equiaxed grains, while increasing the temperature promoted grain growth.

This microstructural evolution is represented in Figure 4.6 by a sequence of SEM images and schematic illustrations. The decrease in the bilayer thickness led to a decrease in the temperature of the exothermic reaction (from 250°C for 14 nm to 200°C for 5 nm) and of the heat of reaction (–364.1 J/g for 14 nm and –246.5 J/g for 5 nm). It should be noted that the amount of heat released by the multilayers with 14 nm of bilayer thickness was still far below that expected, –739 J/g (Morris et al. 2010). The decrease in the heat release resulting from a reduction in the bilayer thickness can be associated with the largest intermixing detected for the smaller bilayer thickness multilayers, as previously stated. As the bilayer thickness decreases, the intermixing affected a larger volume fraction of the film, leading to a reduction in the material that reacted and of the heat released; the intermixing can lead to a decrease of 50 per cent in the heat of reaction of the multilayers. The formation of an intermetallic phase during deposition also reduced the driving force for subsequent reactions in the multilayer.

For multilayers with bilayer thickness larger than 20 nm, the reaction occurred in two stages: the first one associated with the formation of the $NiAl_3$ phase at 230°C, and the second with the formation of Ni_2Al_3 and

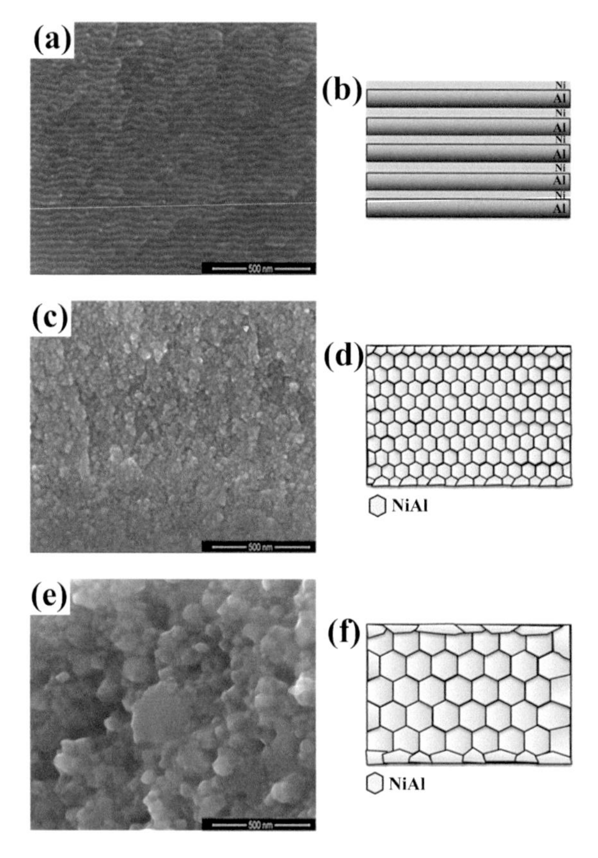

Figure 4.6 SEM images and schematic illustrations of the microstructure evolution of Ni/Al multilayers with bilayer thicknesses smaller than 20 nm: (a) and (b) as deposited (c) and (d) annealed at 400°C and (e) and (f) annealed at 700°C.

NiAl phases at 330°C. However, after heating to 400°C, only NiAl was detected by XRD. The heat released by these reactions was –99.4 J/g for the first and –287.8 J/g for the second. This microstructural evolution is represented schematically in Figure 4.7, including the reaction path and SEM observations.

The as-deposited multilayers and the multilayers heat treated at 250°C and 300°C shown a layered morphology that became fainter as the heat treating temperature rises to 300°C (Figure 4.7 (a) to (e)); this morphology vanished after heat treatment at 400°C (Figure 4.7 (g)). The as-deposited alternated nickel and aluminum layers evolved into nickel and $NiAl_3$ layers. This evolution was responsible for the observed decrease in contrast between adjacent layers on the SEM images. The formation of Ni_2Al_3 must be caused by the reaction of $NiAl_3$ with unreacted nickel. The layered morphology became so faint that it was only visible in some regions of the SEM image. The phases reacted completely at 400°C to form small equiaxed grains of NiAl, visible in the SEM image. Increasing the temperature from 450 to 700°C only promoted grain growth (Figure 4.7).

These results reveal that the bilayer thickness and the chemical composition have a strong influence on the number of reactions of the Ni/Al reactive multilayers produced by sputtering as well as on the first phase to form and the number of intermediate phases.

In summary, the layers of nickel and aluminum of the as-deposited thin films, with equiatomic composition, are transformed into nanocrystalline NiAl grains by annealing, following reaction paths that depend on the bilayer thickness. The increase in the annealing temperature, over the temperature of the NiAl formation, leads to grain growth of the NiAl grains. The final grain size appears to depend on the multilayer bilayer thickness. Another aspect that should be noted is the fact that the heat released by the Ni/Al multilayer reactions is lower than that reported in the literature, as the result of the intermixing that occurs during deposition. This difference increases as the bilayer thickness decreases.

4.2.3 Ni/Ti Reactive Multilayers

Ni/Ti reactive multilayers are a system that was also used as an interlayer in the bonding of similar and dissimilar materials, particularly Ti-based alloys. For γ-TiAl bonding these multilayers have some disadvantages because of the different chemical composition of the base material, unlike the Ti/Al system, and are not as reactive as the Ni/Al reactive multilayers. However, they have proved to be effective in improving bonding of these alloys and other titanium alloys, allowing for a reduction in the bonding conditions in comparison with other multilayers (Simões et al. 2012, 2016b). This is related to the diffusion of nickel and titanium and also to the reaction between these two elements and the base material.

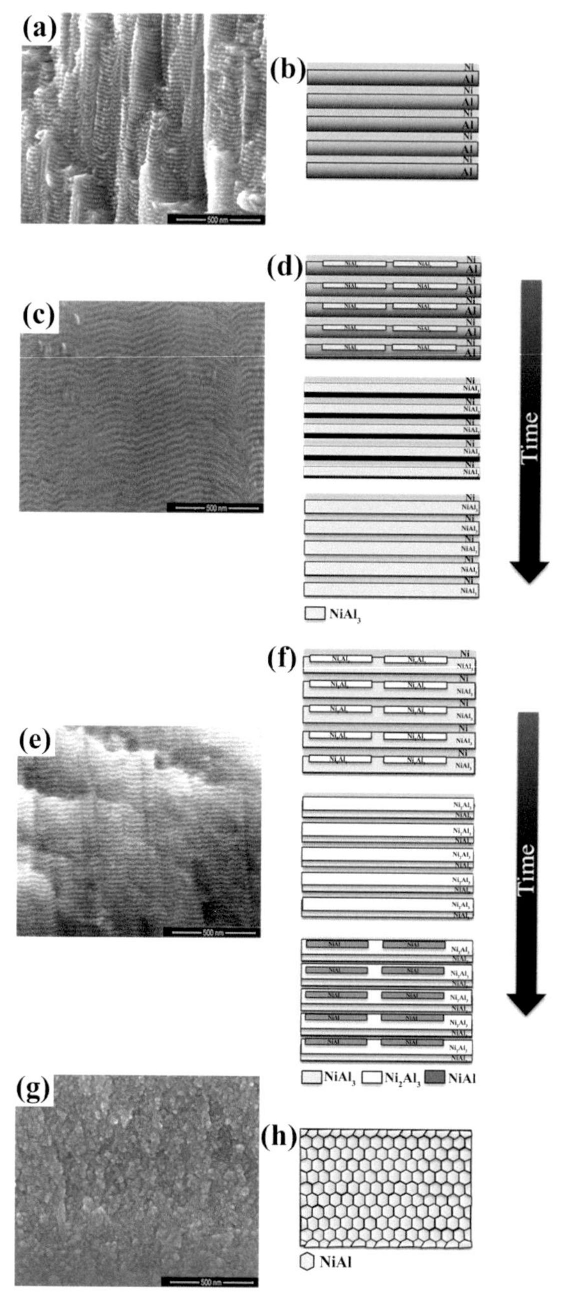

Figure 4.7 SEM images and schematic illustrations of the microstructure evolution of multilayers with bilayer thicknesses higher than 20 nm: (a) and (b) as deposited (c) and (d) annealed at 250°C, (e) and (f) annealed at 300°C and (g) and (h) annealed at 400°C.

As for Ti/Al and Ni/Al systems, Ni/Ti reactive multilayers can be produced with success by magnetron sputtering. The microstructures of as-deposited Ni/Ti reactive multilayers are similar to those already described for Ti/Al and Ni/Al reactive multilayers.

Figure 4.8 shows typical microstructures of Ni/Ti reactive multilayers with bilayer thicknesses of 30 and 60 nm. In this figure the columnar microstructure of the multilayer produced by sputtering can be observed.

Sputter deposition enables easy control of the deposition rate and a flexible adjustment of the chemical composition of the multilayers. Ni/Ti reactive multilayers can be produced with bilayer thicknesses varying between a few nanometers and several micrometers (Bhatt et al. 2006, Cavaleiro et al. 2014, 2015). Ni/Ti reactive multilayers have shown a tendency to present a disordered structure for bilayer thicknesses of a few nanometers. This behaviour, first reported and discussed by Clemens (1986) has been recently reviewed by Cavaleiro et al. (2014, 2015) using HRTEM and synchrotron radiation-based XRD.

For a bilayer thickness of 4 nm, titanium and nickel-rich alternating layers were observed in a TEM image, but it was very difficult to distinguish nickel and titanium peaks in the corresponding Selected Area Electron Diffraction (SAED) patterns. In fact, the SAED pattern showed a diffuse halo, which was indicative of scattering from an amorphous structure. This conclusion was confirmed by grazing incidence XRD. The diffractogram showed only a broad peak around the most intense titanium and nickel diffractions, suggesting a highly disordered or amorphous structure. Multilayers with a bilayer thickness of 12 nm already had a microstructure consisting of nanometric grains of titanium and nickel (Cavaleiro et al. 2015).

For Ni/Ti reactive multilayers, as mentioned for Ni/Al reactive multilayers, intermixing occurred during the sputter deposition. This was restricted to a very localised region. EDS analysis in the TEM of an Ni/Ti reactive multilayer with 28 nm of bilayer thickness showed up to ~ 15 at.% nickel in titanium layers and ~ 14 at.% titanium in nickel layers (Cavaleiro et al. 2015). The intermixing became more pronounced with decreasing bilayer thickness. However, no reaction was observed as a result of this effect.

As described for the previous two systems, the reaction of the Ni/Ti reactive multilayer during annealing has been the subject of various studies (Hollanders et al. 1990, Shen et al. 1993, He and Liu 2006, Adams et al. 2009, Cavaleiro et al. 2014, 2015). These studies analysed different aspects of the reaction, such as the reaction products, the temperature of the reaction, the heat released and the influence of deposition conditions.

Unlike other exothermic multilayers, Ni/Ti with equiatomic composition often react to form multiphases (Adams et al. 2009). Different intermetallic phases can result from the exothermic reaction, such as B2-NiTi, B19′-NiTi, $NiTi_2$, Ni_3Ti and Ni_4Ti_3. At higher temperatures, the formation of oxides should also be considered.

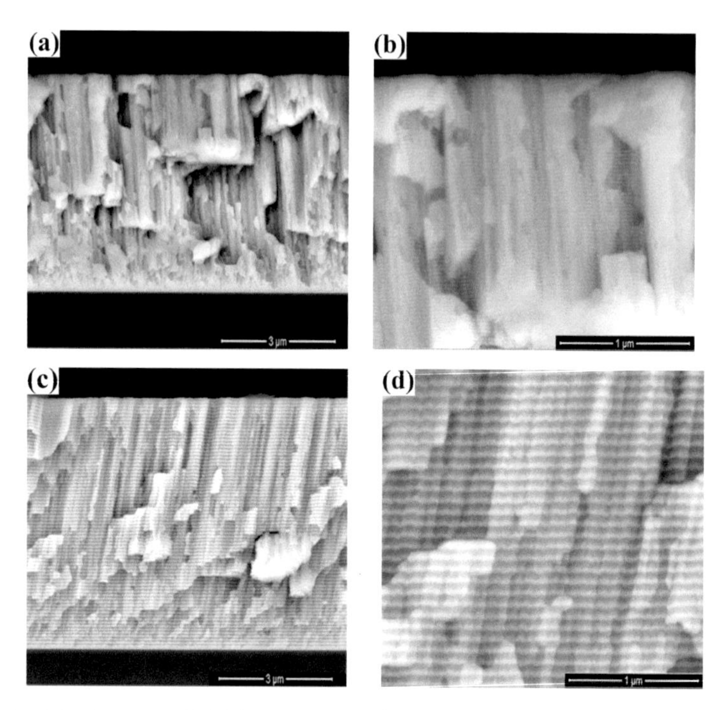

Figure 4.8 SEM images of Ni/Ti reactive multilayers produced by sputtering with bilayer thickness of: (a) and (b) 30 nm and (c) and (d) 60 nm.

Hollanders et al. (1990) showed that diffusion-induced amorphisation can occur at lower annealing temperatures. In fact, this amorphisation occurred at Ni/Ti interfaces and grain boundaries when Ti-40Ni at.% multilayers with 24 nm bilayer thickness were annealed at 250°C. The amorphisation was also observed in the first stage of the annealing of Ni/Ti reactive multilayers produced by cold rolling (Shen et al. 1993). However, this amorphous phase was not detected in near equiatomic multilayers with bilayer thickness between 4 and 25 nm (Cavaleiro et al. 2015).

Table 4.3 shows the temperature of reaction and phase formation in Ni/Ti multilayers with different bilayer thicknesses annealed with and without substrate. The temperature of B2-NiTi formation is strongly related to the bilayer thickness—the increase in bilayer thickness decreases this temperature. This effect was attributed to higher amorphisation observed for smaller bilayer thicknesses, which slows the interdiffusion and thus delays the reaction (Gupta et al. 2006).

The results presented in Table 4.3 show the effect of substrate in the exothermic reaction products and the reaction temperature. However, this last effect was difficult to analyse since different thermal cycles were applied during annealing. The initial reaction product was mainly determined by the multilayer chemical composition. In fact, in multilayers produced with 4, 12, 25 and 70 nm bilayer thickness, a chemical composition of 45 to 50 at.% nickel and freestanding or deposited onto different substrates (stainless steel, austenitic NiTi, Ti6Al4V and silicon), the first phase formed is always B2-NiTi (Cavaleiro et al. 2014, 2015). The layered structure of as-deposited multilayers vanished during annealing when the B2-NiTi equiaxed grains are being formed as can be seen in Figure 4.9.

The substrate chemical composition had a strong influence when the multilayers were annealed at higher temperatures (above 500°C). For the multilayers deposited onto a substrate which did not have titanium, such as stainless steel or silicon, the increase in temperature above the B2-NiTi formation resulted in the formation of titanium dioxide. This causes a nickel enrichment leading to the formation of Ni_3Ti (Cavaleiro et al. 2014).

A similar effect was observed during annealing of freestanding multilayers. On the other hand, if the multilayers were deposited onto a titanium-rich substrate (such as γ-TiAl, Ti6Al4V or NiTi), the formation of oxides was prevented during the heat treatment (Cavaleiro et al. 2015). This can be explained by the high diffusion coefficient of titanium through the nanometric B2-NiTi grains (He and Liu 2006). Titanium diffuses through the B2-NiTi, forming $NiTi_2$ and avoiding the formation of oxide. The structure

Table 4.3 Temperature of reactions and phase formation of 45–50 at.% Ni/Ti reactive multilayers with different bilayer thicknesses deposited in different substrates.

Bilayer thickness (nm)	Temperature of reaction (°C)	Phases	Substrate
4	385	B2-NiTi	Ti-6Al-4V (Cavaleiro et al. 2015)
	526	B2-NiTi + $NiTi_2$	
12	350	B2-NiTi	
	526	B2-NiTi + $NiTi_2$	
25	320	B2-NiTi	
	526	B2-NiTi + $NiTi_2$	
5	400	B2-NiTi	Stainless steel (Cavaleiro et al. 2014)
	600	TiO_2 + Ni_3Ti	
25	375	B2-NiTi	
	600	TiO_2 + Ni_3Ti	
30	444	B2-NiTi	Freestanding multilayers (without substrate)
	728	TiO_2 + Ni_3Ti	
60	414	B2-NiTi	
	690	TiO_2 + Ni_3Ti	

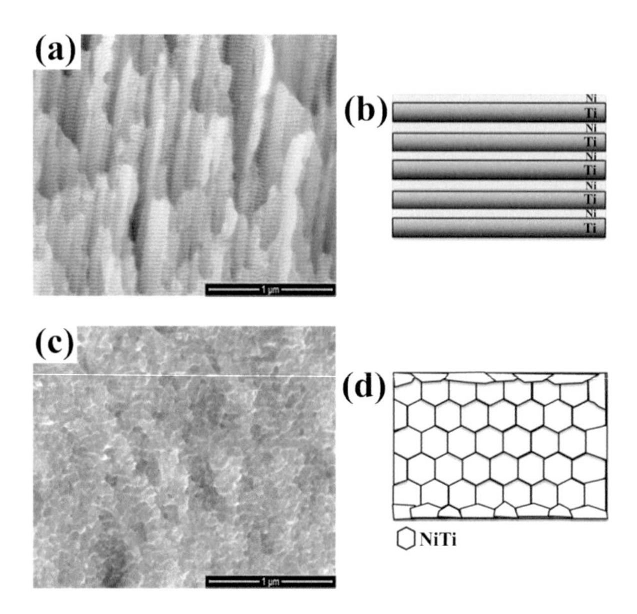

Figure 4.9 SEM images and schematic illustrations of the microstructure evolution of multilayers with bilayer thickness of 30 nm: (a) and (b) as-deposited and (c) and (d) annealed at 500°C.

of a multilayer transformed into B2-NiTi equiaxed grains with nanometric $NiTi_2$ grains is presented in Figure 4.10.

A characteristic of this multilayer was that its surface roughness, which affects the diffusion bonding process, was very sensitive to deposition conditions (Tall et al. 2007). Recently, it was also shown that the bilayer thickness has a similar effect on surface roughness. Cavaleiro et al. (2014) deposited multilayers with bilayer thicknesses of between 5 and 70 nm, only varying the rotation speed of the substrate holder. The morphology of the Ni/Ti reactive multilayer was typical of films produced by sputtering, which exhibited a columnar growth and parallel layers consisting of nanometric grains. The surface roughness was measured by Atomic Force Microscopy (AFM) and was shown to increase with increasing thickness of the bilayer. However, this roughness was reduced and determined by the size of nanometric grains formed on the surface of the film. Smaller bilayer thicknesses were associated with smaller grain size.

Ni/Ti reactive multilayers can also be produced successfully by ARB, allowing the production of a freestanding thick sheet of nickel and titanium nanometric layers (Emadinia et al. 2016). This occurred because nickel and titanium easily bond to each other by severe plastic deformation and exhibited a similar mechanical response to rolling. Although ARB produces alternating nanometric layers, the thicknesses of these layers are not uniform.

Figure 4.10 TEM images of a Ni/Ti reactive multilayer with 30 nm of bilayer thickness annealed at 900°C showing B2-NiTi grains with nanometric $NiTi_2$ grains.

To sum up, Ni/Ti reactive multilayers, though not as exothermic as Ni/Al reactive multilayers can be used as an interlayer in the joining process, taking advantage of the exothermic reaction and the formation of B2-NiTi nanometric grains, less brittle than Ti-Al or Ni-Al intermetallic phases.

4.3 Concluding Remarks

Reactive multilayers can be used as a localised heat source, improving diffusivity due to the nanometric and energetic character of the reaction. The heat released during the exothermic reaction can be used to decrease the processing conditions of joining. However, the key for the success of the application of these reactive multilayers depends on heat release and phase formation during the reaction, and the diffusion and the reaction between the base materials and multilayers. When the multilayers are deposited onto the base materials, it is also necessary to pay special attention to the adhesion between the multilayers and the base material, since this is crucial for a successful application in the joining processes.

The ignition of the reaction of multilayers may occur by thermal explosion, by self-propagating high temperature synthesis or by thermal annealing. These reactions can become self-propagating if the atomic diffusion and energy release rate are fast enough. Multilayers containing aluminum are the most promising systems for use as an interlayer, since they exhibit higher heat release and higher propagation velocities. However, other systems have also proved to be valuable options. The systems shown to be effective in joining processes, particularly in bonding of Ti-based alloys, are Ti/Al, Ni/Al and Ni/Ti reactive multilayers.

Ti/Al and Ni/Ti reactive multilayers have less exothermic reactions than Ni/Al reactive multilayers. This does not imply that these systems are not a good option for use as an interlayer. Ti/Al reactive multilayers are particularly useful in the joining of γ-TiAl alloys. In fact, they have the advantage of having the same chemical composition as the base material and thus ensure microstructural continuity during the bonding process. However, the formation of small amounts of α_2-Ti_3Al, occurring very often during the reaction of these multilayers, can introduce a negative effect on the mechanical properties of the joints. Ni/Ti reactive multilayers can promote an effective bond due to the high titanium diffusion coefficient in NiTi. These multilayers have better characteristics for the bonding of base materials rich in titanium. They also have some advantages due to the more ductile behaviour of the B2-NiTi phase. However, their use for joining materials for high-temperature applications is questionable.

Keywords: ARB; bilayer thickness; diffusion bonding; exothermic reaction; heat released; intermixing; joining technologies; multilayers; Ni/Al multilayers; Ni/Ti multilayers; PVD; sputtering; Ti/Al multilayers.

4.4 References

Adams, D.P., M.A. Rodriguez, J.P. McDonald, M.M. Bai, E. Jones Jr., L. Brewer et al. 2009. Reactive Ni/Ti nanolaminates. J. Appl. Phys. 106: 093505-1-8.

Adams, D.P. 2015. Reactive multilayers fabricated by vapor deposition: A critical review. Thin Solid Films 576: 98–128.

Banerjee, R., X.-D. Zhang, S.A. Dregia and H.L. Fraser. 1999. Phase stability in Al/Ti multilayers. Acta Mater. 47: 1153–1161.

Barmak, K., C. Michaelsen and G. Lucadamo. 1997. Reactive phase formation in sputter-deposited Ni/Al multilayer thin films. J. Mater. Res. 12: 133–146.

Bhatt, P., V. Ganeshan, V.R. Reddy and S.M. Chaudhari. 2006. High temperature annealing effect on structural and magnetic properties of Ti/Ni multilayers. Appl. Surf. Sci. 253: 2572–2580.

Blobaum, K.J., D. Van Heerden, A.J. Gavens and T.P. Weihs. 2003. Al/Ni formation reactions: characterization of the metastable Al9Ni2 phase and analysis of its formation. Acta Mater. 51: 3871–3884.

Boettge, B., J. Braeuer, M. Wiemer, M. Petzold, J. Bagdahn and T. Gessner. 2010. Fabrication and characterization of reactive nanoscale multilayer systems for low-temperature bonding in microsystem technology. J. Micromech. Microeng. 20: 064018-1-8.

Cao, J., X.G. Song, L.Z. Wu, J.L. Qi and J.C. Feng. 2012. Characterization of Al/Ni multilayers and their application in diffusion bonding of TiAl to TiC cermet. Thin Solid Films 520: 3528–3531.

Cavaleiro, A.J., R.J. Santos, A.S. Ramos and M.T. Vieira. 2014. *In situ* thermal evolution of Ni/Ti multilayer thin films. Intermetallics 51: 11–17.

Cavaleiro, A.J., A.S. Ramos, R.M.S. Martins, F.M. Braz Fernandes, J. Morgiel, C. Baehtz et al. 2015. Phase transformations in Ni/Ti multilayers investigated by synchrotron radiation-based x-ray diffraction. J. Alloy. Compd. 646: 1165–1171.

Chen, T., Z.L. Wu, B.S. Cao, J. Gao and M.K. Lei. 2007. Solid state reaction of Fe/Ti nanometer-scale multilayers. Surf. Coat. Tech. 201: 5059–5062.

Clemens, B.M. 1986. Solid-state reaction and structure in compositionally modulated zirconium-nickel and titanium-nickel films. Phys. Rev. B 33: 7615–7624.

Colgan, E.G., M. Nastasi and J.W. Mayer. 1995. Initial phase formation and dissociation in the thin-film Ni/Al system. J. Appl. Phys. 58: 4125–4129.
da Silva Bassani, M.H., J.H. Perepezko, A.S. Edelstein and R.K. Everett. 1997. Initial phase evolution during interdiffusion reactions. Scr. Mater. 37: 227–232.
Ding, H.-S., J.-M. Lee, B.-R. Lee, S.-B. Kang and T.-H. Nam. 2005. Processing and microstructure of TiNi SMA strips prepared by cold roll-bonding and annealing of multilayer. Mater. Sci. Eng. A-Struct. Mater. Prop. Microstruct. Process. 408: 182–189.
Duarte, L.I., A.S. Ramos, M.F. Vieira, F. Viana, M.T. Vieira and M. Koçak. 2006. Solid-state diffusion bonding of gamma-TiAl alloys using Ti/Al thin films as interlayers. Intermetallics 14: 1151–1156.
Edelstein, A.S., R.K. Everett, G.Y. Richardson, S.B. Qadri, E.I. Altman, J.C. Foley et al. 1994. Intermetallic phase formation during annealing of Al/Ni multilayers. J. Appl. Phys. 76: 7850–7859.
Edelstein, A.S., R.K. Everett, G.R. Richardson, S.B. Qadri, J.C. Foley and J.H. Perepezko. 1995. Reaction kinetics and biasing in Al/Ni multilayers. Mater. Sci. Eng. A-Struct. Mater. Prop. Microstruct. Process. 195: 13–19.
Emadinia, O., S. Simões, F. Viana, M.F. Vieira, A.J. Cavaleiro, A.S. Ramos et al. 2016. Cold rolled versus sputtered Ni/Ti multilayers for reaction-assisted diffusion bonding. Weld. World. 60: 337–344.
Gachon, J.-C., A.S. Rogachev, A.E. Grigoryan, H.V. Illarionova, J.-J. Kuntz, D. Yu. Kovalev et al. 2005. On the mechanism of heterogeneous reaction and phase formation in Ti/Al multilayer nanofilms. Acta Mater. 53: 1225–1231.
Gavens, A.J., D. Van Heerden, A.B. Mann, M.E. Reiss and T.P. Weihs. 2000. Effect of intermixing on self-propagating exothermic reaction in Al/Ni nanolaminate foils. J. Appl. Phys. 87: 1255–1263.
Ghalandari, L., M.M. Mahdavian, M. Reihanian and M. Mahmoudiniya. 2016. Production of Al/Sn multilayer composite by accumulative roll bonding (ARB): A study of microstructure and mechanical properties. Mater. Sci. Eng. A-Struct. Mater. Prop. Microstruct. Process. 661: 179–186.
Gupta, R., M. Gupta, S.K. Kulkami, S. Kharrazi, A. Gupta and S.M. Chaudhari. 2006. Thermal stability of nanometer range Ti/Ni multilayers. Thin Solid Films 515: 2213–2219.
He, P. and D. Liu. 2006. Mechanism of forming interfacial intermetallic compounds at interface for solid state diffusion bonding of dissimilar materials. Mater. Sci. Eng. A-Struct. Mater. Prop. Microstruct. Process. 437: 430–435.
Hollanders, M.A., B.J. Thijsse and E.J. Mittemeijer. 1990. Amorphization along interfaces and grain boundaries in polycrystalline multilayers: An x-ray-diffraction study of Ni/Ti multilayers. Phys. Rev. B 42: 5481–5496.
Illeková, E., J.-C. Gachon, A. Rogachev, H. Grigoryan, J.C. Schuster, A. Nosyrev et al. 2008. Kinetics of intermetallic phase formation in the Ti/Al multilayers. Thermochim. Acta 469: 77–85.
Jamaati, R. and M.R. Toroghinejad. 2011. Cold rolling bond strengths: review. Mater. Sci. Tech. Ser. 27: 1101–1108.
Jeske, T., M. Seibt and G. Schmitz. 2003. Microstructural influence on the early stages of interreaction of Al/Ni-investigated by TAP and HREM. Mater. Sci. Eng. A-Struct. Mater. Prop. Microstruct. Process. 353: 105–111.
Lee, S.-G., S.-P. Kim, K.-R. Lee and Y.-C. Chung. 2005. Atomic-level investigation of interface structure in Ni-Al multilayer system: molecular dynamics simulation. J. Magn. Magn. Mater. 286: 394–398.
Lucadamo, G., K. Barmak, S. Hyun, C. Cabral Jr. and C. Lavoie. 1999. Evidence of a two-stage reaction mechanism in sputter deposited Nb/Al multilayer thin-films studied by in situ synchrotron X-ray diffraction. Mater. Lett. 39: 268–273.
Lucadamo, G., K. Barmak, D.T. Carpenter and J.M. Rickman. 2001. Microstructure evolution during solid state reactions of Nb/Al multilayers. Acta Mater. 49: 2813–2826.

Ma, E., M.-A. Nicolet and M. Nathan. 1989. NiAl3 formation in Al/Ni thin-film bilayers with and without contamination. J. Appl. Phys. 65: 2703–2710.
Ma, E., C.V. Thompson, L.A. Clevenger and K.N. Tu. 1990. Self-propagating explosive reactions in Al/Ni multilayer thin films. Appl. Phys. Lett. 57: 1262–1264.
Ma, E., C.V. Thompson and L.A. Clevenger. 1991. Nucleation and growth during reactions in multilayer Al/Ni films: the early stage of Al3Ni formation. J. Appl. Phys. 69: 2211–2218.
Michaelsen, C., S. Wöhlert and R. Bormann. 1994. Phase formation and microstructural development during solid-state reactions in Ti-Al multilayer films. Mater. Res. Soc. Symp. Proc. 343: 205–210.
Michaelsen, C., S. Wöhlert, R. Bormann and K. Barmak. 1995. The early stages of solid-state reactions in Ti/Al multilayer films. Mater. Res. Soc. Symp. Proc. 398: 245–250.
Michaelsen, C., G. Lucadamo and K. Barmak. 1996. The early stages of solid-state reactions in Ni/Al multilayer films. J. Appl. Phys. 80: 6689–6698.
Morris, C.J., B. Mary, E. Zakar, S. Barron, G. Fritz, O. Knio et al. 2010. Rapid initiation of reactions in Al/Ni multilayers with nanoscale layering. J. Phys. Chem. Solids 71: 84–89.
Murphy, R.D., R.V. Reeves, C.D. Yarrington and D.P. Adams. 2015. The dynamics of Al/Pt reactive multilayer ignition via pulsed-laser irradiation. Appl. Phys. Lett. 107: 234103-1-5.
Nathani, H., J. Wang and T.P. Weihs. 2007. Long-term stability of nanostructured systems with negative heats of mixing. J. Appl. Phys. 101: 104315-1-4.
Noro, J., A.S. Ramos and M.T. Vieira. 2008. Intermetallic phase formation in nanometric Ni/Al multilayer thin films. Intermetallics 16: 1061–1065.
Politano, O., F. Baras, A.S. Mukasyan, S.G. Vadchenko and A.S. Rogachev. 2013. Microstructure development during NiAl intermetallic synthesis in reactive Ni-Al nanolayers: Numerical investigations vs. TEM observations. Surf. Coat. Tech. 215: 485–492.
Qiu, X. and J. Wang. 2007. Experimental evidence of two-stage formation of Al3Ni in reactive Ni/Al multilayer foils. Scr. Mater. 56: 1055–1058.
Ramos, A.S. and M.T. Vieira. 2005. Kinetics of the thin films transformation Ti/Al multilayer → γ-TiAl. Surf. Coat. Tech. 200: 326–329.
Ramos, A.S. and M.T. Vieira. 2012. Intermetallic compound formation in Pd/Al multilayer thin films. Intermetallics 25: 70–74.
Ramos, A.S., R. Calinas and M.T. Vieira. 2006. The formation of γ-TiAl from Ti/Al multilayers with different periods. Surf. Coat. Tech. 200: 6196–6200.
Ramos, A.S., M.T. Vieira, J. Morgiel, J. Grzonka, S. Simões and M.F. Vieira. 2009. Production of intermetallic compounds from Ti/Al and Ni/Al multilayer thin films—A comparative study. J. Alloy. Compd. 484: 335–340.
Ramos, A.S., A.J. Cavaleiro, M.T. Vieira, J. Morgiel and G. Safran. 2014. Thermal stability of nanoscale metallic multilayers. Thin Solid Films 571: 268–274.
Rogachev, A.S., A.É. Grigoryan, E.V. Illarionova, I.G. Kanel, A.G. Merzhanov, A.N. Nosyrev et al. 2004. Gasless combustion of Ti-Al bimetallic multilayer nanofoils. Combust. Explos. 40: 166–171.
Rogachev, A.S. 2008. Exothermic reaction waves in multilayer nanofilms. Russ. Chem. Rev. 77: 21–37.
Rogachev, A.S., S.G. Vadchenko, F. Baras, O. Politano, S. Rouvimov, N.V. Sachkova et al. 2014. Structure evolution and reaction mechanism in the Ni/Al reactive multilayer nanofoils. Acta Mater. 66: 86–96.
Rogachev, A.S., S.G. Vadchenko, F. Baras, O. Politano, S. Rouvimov, N.V. Sachkova et al. 2016. Combustion in reactive multilayer Ni/Al nanofoils: Experiments and molecular dynamic simulation. Combust. Flame 166: 158–169.
Shen, T.D., M.X. Quan and J.T. Wang. 1993. Solid state amorphization reactions in Ni/Ti multilayer composites prepared by cold rolling. J. Mater Sci 28: 394–398.
Sieber, H., J.S. Park, J. Weissmüller and J.H. Perepezko. 2001. Structural evolution and phase formation in cold-rolled aluminum-nickel multilayers. Acta Mater. 49: 1139–1151.
Simões, S., F. Viana, A.S. Ramos, M.T. Vieira and M.F. Vieira. 2010a. TEM characterization of as-deposited and annealed Ni/Al multilayer thin film. Microsc. Microanal. 16: 662–669.

Simões, S., F. Viana, V. Ventzke, M. Koçak, A.S. Ramos, M.T. Vieira et al. 2010b. Diffusion bonding of TiAl using Ni/Al multilayers. J. Mater. Sci. 45: 4351–4357.
Simões, S., F. Viana, A.S. Ramos, M.T. Vieira and M.F. Vieira. 2011. Anisothermal solid-state reactions of Ni/Al nanometric multilayers. Intermetallics 19: 350–356.
Simões, S., F. Viana, A.S. Ramos, M.T. Vieira and M.F. Vieira. 2013. Reaction zone formed during diffusion bonding of TiNi to Ti6Al4V using Ni/Ti nanolayers. J. Mater. Sci. 48: 7718–7727.
Simões, S., F. Viana and M.F. Vieira. 2014. Reactive commercial Ni/Al nanolayers for joining lightweight alloys. J. Mater. Eng. Perform. 23: 1536–1543.
Simões, S., A.S. Ramos, F. Viana, O. Emadinia, M.T. Vieira and M.F. Vieira. 2016a. Ni/Al multilayers produced by accumulative roll bonding and sputtering. J. Mater. Eng. Perform. 25: 4394–4401.
Simões, S., F. Viana, A.S. Ramos, M.T. Vieira and M.F. Vieira. 2016b. Microstructural characterization of diffusion bonds assisted by Ni/Ti nanolayers. J. Mater. Eng. Perform. 23: 3245–3251.
Simões, S., A.S. Ramos, F. Viana, M.T. Vieira and M.F. Vieira. 2016c. Joining of TiAl to steel by diffusion bonding with Ni/Ti reactive multilayers. Metals 6: 96-1-11.
Simões, S., F. Viana, A.S. Ramos, M.T. Vieira and M.F. Vieira. 2016d. Reaction-assisted diffusion bonding of TiAl alloy to steel. Mater. Chem. Phys. 171: 73–82.
Sun, Y.-B., Y.-Q. Zhao, D. Zhang, C.-Y Liu, H.-Y. Diao and C.-L. Ma. 2011. Multilayered Ti-Al intermetallic sheets fabricated by cold rolling and annealing of titanium and aluminum foils. Trans. Nonferrous Met. Soc. China 21: 1722–1727.
Swiston Jr., A.J., T.C. Hufnagel and T.P. Weihs. 2003. Joining bulk metallic glass using reactive multilayer foils. Scr. Mater. 48: 1575–1580.
Swiston Jr., A.J., E. Besnoin, A. Duckham, O.M. Knio, T.P. Weihs and T.C. Hufnagel. 2005. Thermal and microstructural effects of welding metallic glasses by self-propagating reactions in multilayer foils. Acta Mater. 53: 3713–3719.
Tall, P.D., S. Ndiaye, A.C. Beye, Z. Zong, W.O. Soboyejo, H.-J. Lee et al. 2007. Nanoindentation of Ni-Ti thin films. Mater. Manuf. Process. 22: 175–179.
Trenkle, J.C., J. Wang, T.P. Weihs and T.C. Hufnagel. 2005. Microstructural study of an oscillatory formation reaction in nanostructured reactive multilayer foils. Appl. Phys. Lett. 87: 153108–1-3.
Trenkle, J.C., T.P. Weihs and T.C. Hufnagel. 2008. Fracture toughness of bulk metallic glass welds made using nanostructured reactive multilayer foils. Scr. Mater. 58: 315–318.
Trenkle, J.C., L.J. Koerner, M.W. Tate, N. Walker, S.M. Gruner, T.P. Weihs et al. 2010. Time-resolved x-ray microdiffraction studies of phase transformations during rapidly propagating reactions in Al/Ni and Zr/Ni multilayer foils. J. Appl. Phys. 107: 113511-1-12.
Ustinov, A.I., Yu.I. Falchenko, A.Ya. Ishchenko, G.K. Kharchenko, Y.V. Melnichenko et al. 2008. Diffusion bonding of γ-TiAl based alloys through nano-layered foil of Ti/Al system. Intermetallic 16: 1043–1045.
Weihs, T.P. 2014. Fabrication and characterization of reactive multilayer films and foils. pp. 160–243. *In*: K. Barmak and K. Coffey [eds.]. Metallic Films for Electronic, Optical and Magnetic Applications. Woodhead Publishing Limited, Cambridge, England, UK.
Woll, K., A. Bergamaschi, K. Avchachov, F. Djurabekova, S. Gier, C. Pauly et al. 2016. Ru/Al multilayers integrate maximum energy density and ductility for reactive materials. Sci. Rep. 6: 19535; doi:10.1038/srep19535.
Yajid, M.A.M., R.C. Doole, T. Wagner and G. Möbus. 2008. Heating and EELS experiments of CuAl reactive multilayers. J. Phys.: Conf. Ser. 126: 012064-1-4.

CHAPTER 5

Diffusion Bonding of γ-TiAl Alloys Assisted by Reactive Multilayers

5.1 Introduction

Diffusion bonding is one of the most promising processes for joining γ-TiAl alloys. This process allows successful bonding to similar and dissimilar materials with appreciable mechanical properties. However, as already mentioned, the demanding bonding conditions (temperature, time and pressure) necessary to achieve a successful joint makes the process less attractive, particularly for industrial implementation.

New approaches to diffusion bonding have been investigated in order to overcome the need for elevated temperatures and pressures for long times and for Post-Bond Heat Treatments (PBHT). Interlayers can be used in the diffusion bonding of the γ-TiAl alloys with the aim of improving contact between the mating surfaces and reducing the formation of brittle compounds. Some investigations have been carried out in the production of these bonds with interlayers, such as sheets, thin foils or thin films. These studies demonstrate that the use of such interlayers does not lead to significant improvements, and that it is usually necessary to implement PBHT at high temperatures in order to obtain the desired mechanical properties.

Reactive multilayers began to be used to obtain or to improve the bond between materials that are difficult to join by conventional processes, such as metallic glasses, electronic devices and metals to ceramics. The successful application of reactive multilayers drove several investigations into their use for bonding other challenging materials or for those requiring very demanding processing conditions, such as γ-TiAl alloys. The multilayers can be deposited on the base materials or can be used as a freestanding film. Their use in diffusion bonding induces a reduction in temperature, pressure and bonding time, since they act as an additional heat source, located at the interface. In addition, these interlayers enhance the contact between the mating surfaces and improve diffusion, due to the nanometric grain size and nanometric thickness of the layers. In brazing processes, the heat released by the reaction can be sufficient to melt the brazing filler without the need for an external heat source.

The selection of multilayer systems for application in diffusion bonding is based not only on the heat released by the reaction, but also on the chemical composition of the base materials in order to avoid the formation of brittle phases that will affect the mechanical properties of the joints. One strategy is the selection of multilayers either with a chemical composition very similar to that of the materials to be bonded or that do not form brittle compounds at the interface. However, multilayer systems that do not meet these requirements may also be used to improve the diffusion bonding of the γ-TiAl alloys.

Diffusion bonding of γ-TiAl alloys can be improved using Ti/Al, Ni/Al and Ni/Ti reactive multilayers. The characteristics and advantages of each of the systems, evaluated by the formation of an interface free from defects and with appreciable mechanical properties, will be discussed in the following sub chapters. The improvement of the process, by diminishing the bonding conditions, without impairing the mechanical properties of the joints, will also be discussed, and a comparison will be made with conventional processes.

The diffusion bonding of γ-TiAl alloys to steel and superalloys using reactive multilayers will be discussed in Chapters 6 and 7, and joining with the combination of reactive multilayers and brazing alloys will be the subject of Chapter 8.

5.2 Diffusion Bonding of γ-TiAl Alloys using Ti/Al Reactive Multilayers

Ti/Al reactive multilayers can be applied to improve the diffusion bonding of the γ-TiAl alloys. The high diffusivity of these nanoscale multilayers improves bonding without introducing chemical or structural discontinuities into the interface (Duarte et al. 2006a,b, Ramos et al. 2006, Ustinov et al. 2008).

Ustinov et al. (2008) investigated the use of Ti/Al reactive multilayers deposited by electron beam in the diffusion bonding of a γ-TiAl alloy (Ti-48Al at.%). The bonding experiments were performed at temperatures ranging from 900 to 1,200°C, with bonding times between 5 and 25 min and pressures from 10 to 70 MPa, in a vacuum chamber. The use of Ti/Al reactive multilayers with 20 μm thickness and 50 nm bilayer thickness led to an improvement in the diffusion bonding of γ-TiAl alloy.

Bonding without the multilayers at 1,200°C for 20 min under a pressure of 70 MPa resulted in an interface with a visible bond line, due to the formation of a α_2-Ti_3Al interlayer, Figure 5.1. While the use of Ti/Al reactive multilayers enabled the production of a sound interface at the same bonding temperature and the same dwell time, but with a reduced pressure of 10 MPa, the bond line disappeared and an interface free from defects was formed, consisting of small γ-TiAl grains, Figure 5.2.

The hardness evaluation revealed an interface with a hardness value similar to the base material. Some of the elements from the base material were observed at the interface, attesting to the diffusion during the joining.

Duarte et al. and Ramos et al. (Duarte et al. 2006a,b, Ramos et al. 2006) demonstrated that Ti/Al reactive multilayers could be used to improve diffusion bonding of γ-TiAl alloy, thus reducing the bonding conditions. The reactive multilayers were deposited onto the mating surfaces of a Ti-45Al-2Cr-2Nb at.% alloy by magnetron sputtering with bilayer

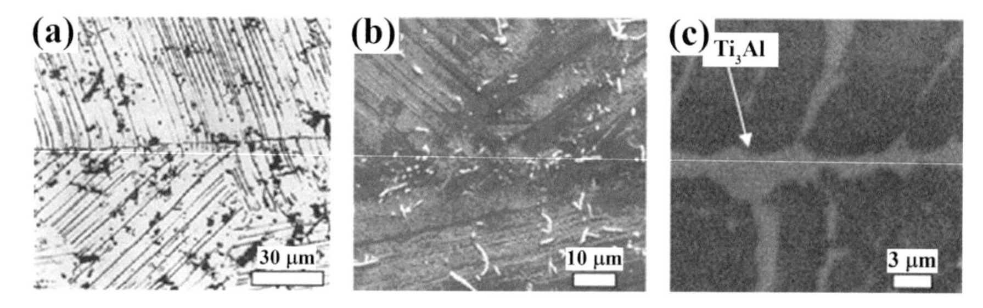

Figure 5.1 Scanning Electron Microscopy (SEM) images of the interface produced without multilayers at 1,200°C under a pressure of 70 MPa for 20 min. (Reprinted from Intermetallics 16: 1043–1045. Ustinov, A.I., Yu.V. Falchenko, A.Ya. Ishchenko, G.K. Kharchenko, T.V. Melnichenko and A.N. Muraveynik. 2008. Diffusion welding of γ-TiAl based alloys through nanolayered foil of Ti/Al system. Copyright 2008, with permission from Elsevier.)

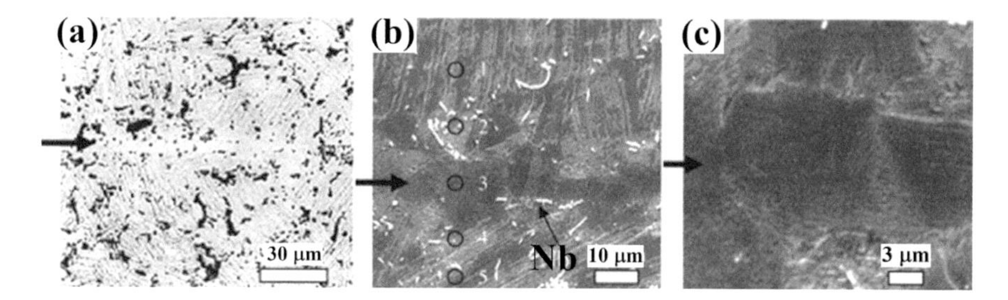

Figure 5.2 SEM images of the interface produced with Ti/Al reactive multilayers at 1,200°C for 20 min under a pressure of 10 MPa. (Reprinted from Intermetallics 16: 1043–1045. Ustinov, A.I., Yu.V. Falchenko, A.Ya. Ishchenko, G.K. Kharchenko, T.V. Melnichenko and A.N. Muraveynik. 2008. Diffusion welding of γ-TiAl based alloys through nanolayered foil of Ti/Al system. Copyright 2008, with permission from Elsevier.)

thicknesses of 4 and 20 nm. These multilayers were deposited with an overall chemical composition close to Ti-49Al at.% and a thickness of 2.0–2.5 μm. The joining experiments were performed at temperatures ranging from 600 to 1,000°C under a pressure of 50 MPa for 60 min.

To improve bonding process, the multilayers deposition began and ended with a titanium layer. The results revealed that a low bonding temperature, 600 to 800°C, produced joints with a lack of union at the edges of the interface and porosity. A thick and non-uniform titanium-rich layer was formed close to the γ-TiAl and at the center of the interface. These layers resulted from the fact that the two mating layers were composed of titanium. Diffusion bonding at higher temperatures using multilayers with 4 nm bilayer thickness without the thicker titanium layers (to eliminate the chemical discontinuities at the interface) were successful, reducing the unbonded regions and porosity.

At 900°C sound joints were produced (Figure 5.3) without the need for titanium thicker layers deposited at the surface of the multilayers, although residual porosities remained at the interface. The increase in the bonding temperature to 1,000°C led to the formation of a uniform interface, free from defects and chemical discontinuities (Figure 5.4). The joint interface was mainly composed of ultra-fine and nanometric grains of γ-TiAl, and some α_2-Ti_3Al grains were also observed.

Although the use of reactive Ti/Al reactive multilayers allowed the diffusion bonding temperature of γ-TiAl alloy (Ti-45Al-2Cr-2Nb at.%) to be reduced, it was still too high to make this approach attractive. For this reason, Duarte et al. (2008, 2012) suggested the use of Ti/Al multilayers doped with silver or copper. Silver and copper were added to promote the

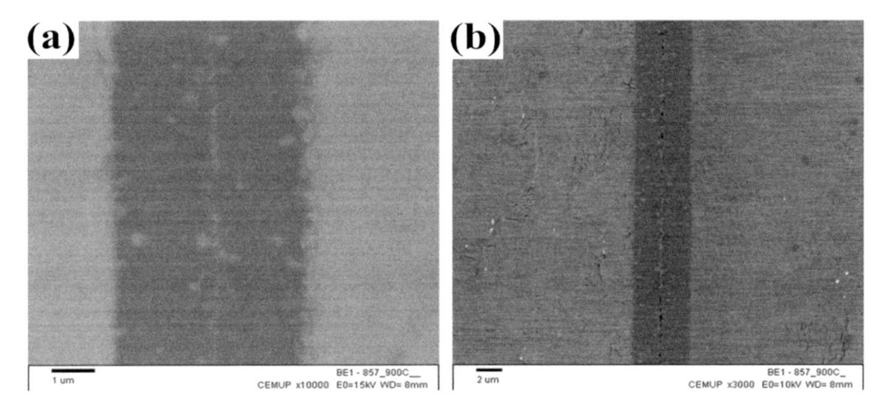

Figure 5.3 SEM images of the interface produced with Ti/Al reactive multilayers with 4 nm bilayer thickness at 900°C for 60 min under a pressure of 50 MPa. (Reprinted from Intermetallics 14: 1151–1156. Duarte, L.I., A.S. Ramos, M.F. Vieira, F. Viana, M.T. Vieira and M. Koçak. 2006. Solid-state diffusion bonding of gamma-TiAl alloys using Ti/Al thin films as interlayers. Copyright 2006, with permission from Elsevier.)

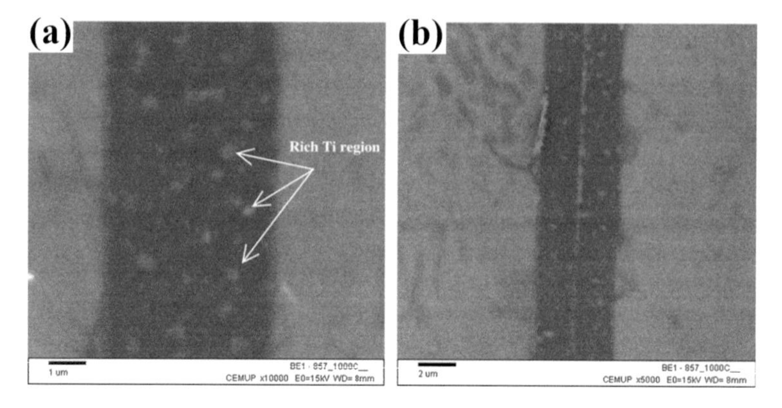

Figure 5.4 SEM images of the interface produced with Ti/Al multilayers with 4 nm bilayer thickness at 1,000°C for 60 min under a pressure of 50 MPa. (Reprinted from Intermetallics 14: 1151–1156. Duarte, L.I., A.S. Ramos, M.F. Vieira, F. Viana, M.T. Vieira and M. Koçak. 2006. Solid-state diffusion bonding of gamma-TiAl alloys using Ti/Al thin films as interlayers. Copyright 2006, with permission from Elsevier.)

multilayers' reaction, enhancing diffusivity during diffusion bonding and improving bond strength. Ti/Al multilayers doped with 7 at.% of copper or 2.8 at.% of silver with 4 nm of bilayer thickness were deposited in the γ-TiAl alloy by magnetron sputtering. Diffusion bonding experiments were conducted at 900°C for 60 min and under a pressure of 50 MPa.

Sound joints were produced with both multilayers doped, free from porosities and without unbonded areas. The interfaces produced with the multilayers doped with silver were mainly composed of ultra-fine and nanometric grains of γ-TiAl and small amounts of α_2-Ti_3Al with dissolved silver. The interfaces produced with the multilayers doped with copper were also composed mainly of nanometric grains of γ-TiAl. α_2-Ti_3Al grains were also observed in addition to an AlCuTi phase. Silver and copper additions to the multilayers were responsible for an increase in the hardness of the interface from 7.4 GPa, when using undoped multilayers, to 8.7 and 8.4 GPa when the multilayers were doped with silver and copper, respectively. These results indicated that the addition of small amounts of silver or copper to the Ti/Al multilayers made them more promising interlayers for diffusion bonding of γ-TiAl alloys.

The use of Ti/Al multilayers was shown to be effective in improving the diffusion bonding of γ-TiAl alloys without defects, by lowering the temperature. However, the enhancement of the diffusion bonding process was not very significant. It was expected that the processing conditions could be further reduced with the use of these multilayers. This can be explained by the fact that the Ti/Al multilayer system is not the most reactive system and so the heat released is not sufficient to allow bonding using lower temperatures and pressures. Although the use of Ti/Al reactive multilayers doped with silver or copper has led to the production of sound joints at a temperature below that used in conventional diffusion bonding processes, the required pressure (50 MPa) was very high, which could have had a negative effect when used to bond more ductile γ-TiAl alloys, as was the case with niobium-rich alloys. A solution for the production of sound interfaces with lower pressures consisted in using other multilayer systems such as Ni/Al or Ti/Ni.

5.3 Diffusion Bonding of γ-TiAl Alloys using Ni/Al Reactive Multilayers

Diffusion bonding of γ-TiAl alloys can be improved by the use of an interlayer composed of Ni/Al reactive multilayers. These are more reactive than Ti/Al multilayers, which make their application in the diffusion bonding of γ-TiAl alloys more appealing.

Cao et al. (2008) studied the applicability of the Ni/Al reactive multilayers in the diffusion bonding of γ-TiAl alloy to enhance their joining efficiency. γ-TiAl (Ti-48Al-2Cr-2Nb at.%) mating surfaces were coated with Ni/Al reactive multilayers by magnetron sputtering. The Ni/Al film had a total thickness of 20–30 μm, a bilayer thickness of 1 μm and a chemical composition close to Ni_3Al. The diffusion bonding experiments were performed at temperatures ranging from 700–900°C for 10 min under a pressure ranging from 35 to 55 MPa. The quality of the joints was evaluated by the combination of microstructural characterisation with the shear tests. Joints without unbonded regions at the interfaces were only observed when produced at the higher temperature (900°C).

At this temperature, the interlayer is mainly composed of Ni_3Al with complex Al-Ni-Ti compounds at the interfaces between the γ-TiAl alloy and multilayers. The observation of the fracture surface of the shear tested samples showed that the fracture had occurred mainly through an Al_2NiTi layer formed at these interfaces. Notwithstanding the formation of this brittle phase, the diffusion bonding of the γ-TiAl alloy using Ni/Al multilayers was performed successfully, attaining an average shear strength of 160 MPa. These multilayers ensured a sound joint under slightly less demanding conditions (less time and lower pressure) than those required for diffusion bonding assisted by Ti/Al reactive multilayers. However, the pressure was still very high and the joint shear strength was lower than that of the base material.

High-strength joints produced at lower pressures and bonding temperatures are essential for the success of this approach. The proper choice of the Ni/Al reactive multilayers, optimising bilayer thickness and chemical composition, is the key to improving the reaction-assisted diffusion bonding of γ-TiAl alloys.

The use of multilayers with nanometric bilayer thickness led to an increase in diffusivity, thereby positively affecting the processing conditions. In fact, Ni/Al multilayers with a nanometric bilayer thickness (5 to 30 nm) were shown to be more effective in reducing joining conditions in diffusion bonding of γ-TiAl alloys (Simões et al. 2010, 2012, 2015). In these studies, alternating nickel and aluminum nanolayers were deposited onto γ-TiAl (Ti-45Al-5Nb at.%) base material by magnetron sputtering. The multilayers had a near equiatomic composition, a bilayer thickness of 5, 14 or 30 nm and a total thickness ranging from 2.0 to 3.5 μm. The use of multilayers allowed diffusion bonding of γ-TiAl alloys with pressures of 5 and 10 MPa at temperatures ranging from 800 to 1,000°C.

The bilayer thickness of the multilayer is an important factor influencing bond quality and mechanical strength. At the lower bonding temperature (800°C), only the multilayers with 14 nm of bilayer thickness, processed under a pressure of 10 MPa, enabled the production of a well-bonded and defect free interface (Figure 5.5). This can be explained by the combined effects of intermixing, heat released and intermediate phases formed. Compared with these multilayers, the amount of heat released by the other two multilayers (5 and 30 nm of bilayer thickness) was smaller. In fact, for 5 nm of bilayer thickness, intermixing is greater, due to the larger number of interfaces, and the heat released by the multilayers is lower, while for 30 nm, the heat released is divided into stages due to the formation of intermediate phases before the formation of the final product (NiAl phase).

The interface shown in Figure 5.5 is thin (a thickness of less than 5 μm) and composed of a columnar layer close to the γ-TiAl and a central area

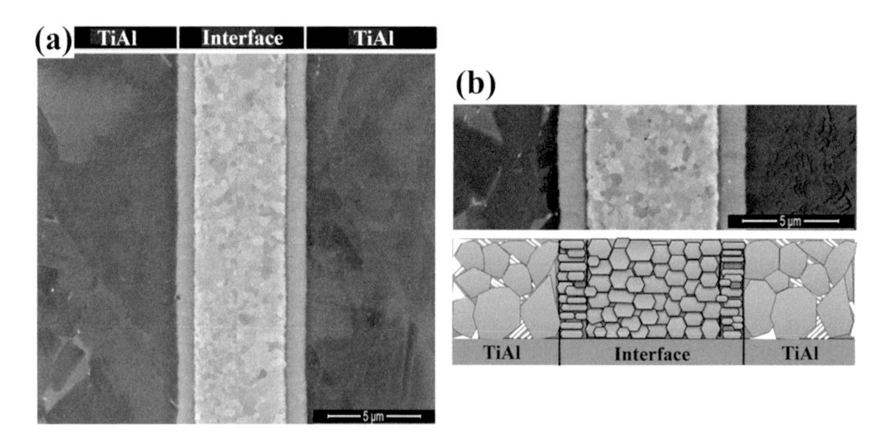

Figure 5.5 (a) SEM image of the bond interface obtained at 800°C for 60 min under a pressure of 10 MPa using Ni/Al reactive multilayers with a bilayer thickness of 14 nm and (b) detail and schematic illustration of the layers that constitute the interface.

with equiaxed grains, which constitute the majority of the interface. Figure 5.5 (b) shows a schematic illustration of the two layers. The microstructure of the γ-TiAl alloy remains unaffected by the bonding process.

Abnormal grain growth is observed in few regions of the central layer of the interface. A slight bond line is observed at the center of the interface, which means that the bonding processing conditions are insufficient to allow intense diffusion across the interface, thus eliminating this line.

A more in-depth analysis of the microstructure of the interface is shown in Figure 5.6. The figure shows Scanning Transmission Electron Microscopy (STEM) dark-field images of the two regions referred to. These images clearly show the columnar structure observed in the layer close to the γ-TiAl alloy and the equiaxed grains in the central region of the interface.

The nanometric and sub micrometer grain size of the phases present at the interface implies that their identification took into consideration the chemical compositions determined by Energy-Dispersive X-Ray Spectroscopy (EDS), the Kikuchi patterns produced by Electron Backscatter Diffraction (EBSD) and diffraction patterns obtained by Selected Area Electron Diffraction (SAED). The results revealed that the central region is composed of NiAl equiaxed grains, while the layer close to the γ-TiAl comprised columnar grains of AlNiTi. The NiAl equiaxed grains were formed due to the multilayer reaction, while the diffusion between the γ-TiAl alloy and the multilayers leads to the formation of the AlNiTi intermetallic compound.

Increasing the bonding temperature to 900°C led to the formation of a sound interface using the Ni/Al reactive multilayers with any of the three-bilayer thicknesses (5, 14 and 30 nm). Figure 5.7 shows the microstructure

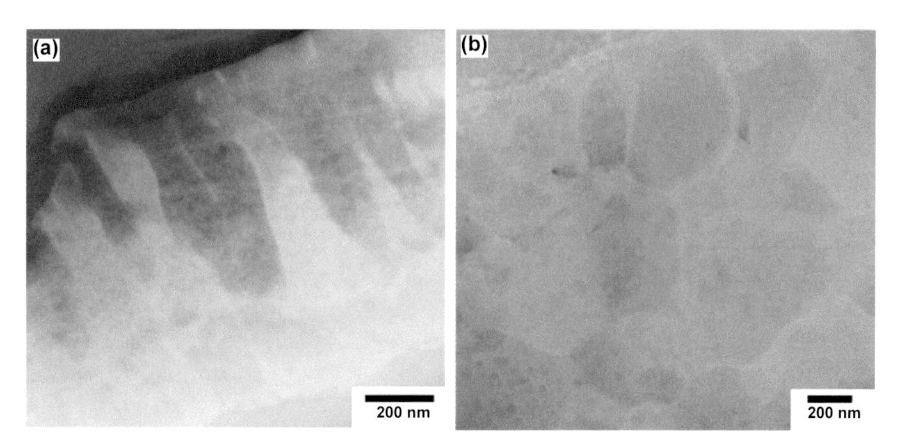

Figure 5.6 STEM dark-field images of (a) the layer close to γ-TiAl base material showing the columnar grains and (b) the central region composed of equiaxed grains.

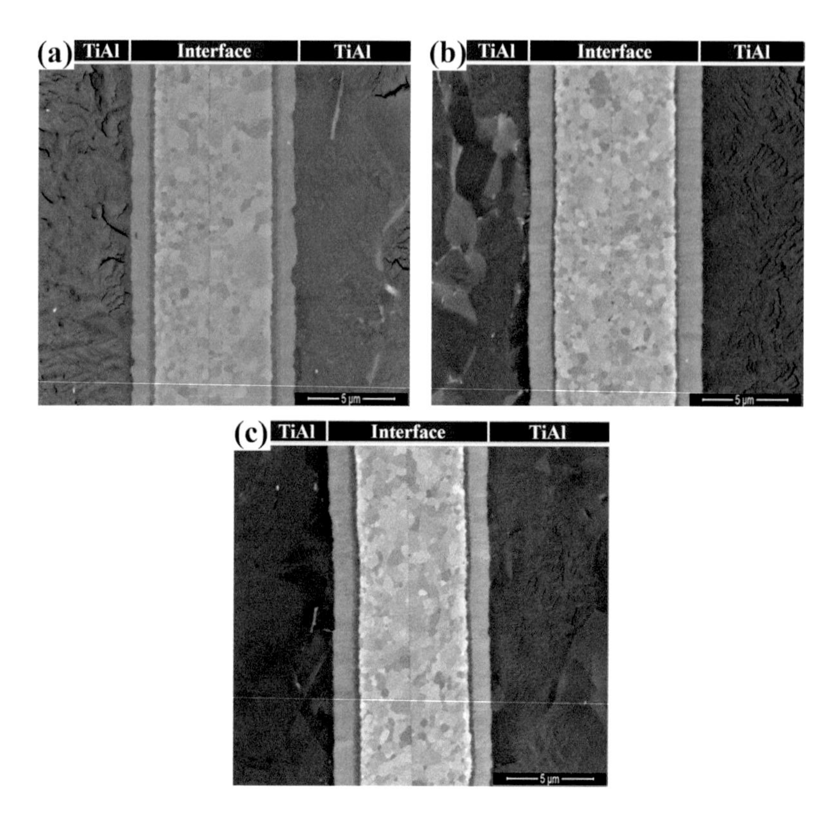

Figure 5.7 SEM images of the bond interface obtained at 900°C for 60 min under a pressure of 5 MPa using Ni/Al reactive multilayers with a bilayer thickness of (a) 30 nm, (b) 14 nm and (c) 5 nm.

of the diffusion bonding interface of γ-TiAl alloy using Ni/Al reactive multilayers, when processed at 900°C for 60 min under a pressure of 5 MPa.

The use of different bilayer thicknesses or the increase in the bonding temperature did not significantly affect the microstructure of the interface. In fact, the interfaces produced with the three bilayer thicknesses are very similar to each other and also to those already described for the temperature of 800°C.

Despite the similarity of the microstructure, the increase in the bonding temperature promoted a more intense diffusion with several effects, two of which were the increase in the interface thickness and in the size of the equiaxed interfacial grains. This more intense diffusion also promoted bonding using multilayers with 5 and 30 nm of bilayer thicknesses. However, a detailed observation shows that a bond line is visible for the interface of the joint produced with the smaller bilayer thickness. For the joints produced with the two other bilayer thicknesses, the bond line had completely disappeared.

The SEM images of Figure 5.7 show a dark line between the layer of columnar grains and the central region. This results from a difference in height between the two regions due to the different response to metallographic preparation of the cross-sections of the phases comprising each region. The presence of different phases at the interface was confirmed by EBSD analysis. EBSD Kikuchi pattern indexation revealed that the columnar grains are from Al_2NiTi and AlNiTi intermetallics and central equiaxed grains are constituted by NiAl phase. The more intense diffusion of aluminum occurring at 900°C promoted the formation of the Al_2NiTi phase in the layer closest to the γ-TiAl alloy.

EDS analyses performed in the TEM also determined the chemical distribution across the interface. Elemental maps revealed a higher concentration of titanium at the center of the interface that was not detected by SEM images and EBSD results. HRTEM observations, combined with Fast Fourier Transform (FFT) and Electron Energy-Loss Spectroscopy (EELS) analysis, revealed that this enrichment in titanium is due to the presence of Ti-rich nanometric grains formed at the bond line.

Figure 5.8 shows the HRTEM image and an EELS map of titanium of this region. The formation of these grains could be due to the decrease in the solubility of titanium on NiAl with the decrease of the temperature that promoted the rejection of titanium for a less compact region (the center of the interface).

The microstructure and the phases identified are very different from those reported by Cao et al. (2008). This can be explained by differences in the Al/Ni atomic ratio and multilayer bilayer thicknesses. In multilayers with a larger amount of nickel and a higher bilayer thickness the reaction occurs in several steps and with the formation of nickel-rich intermetallic

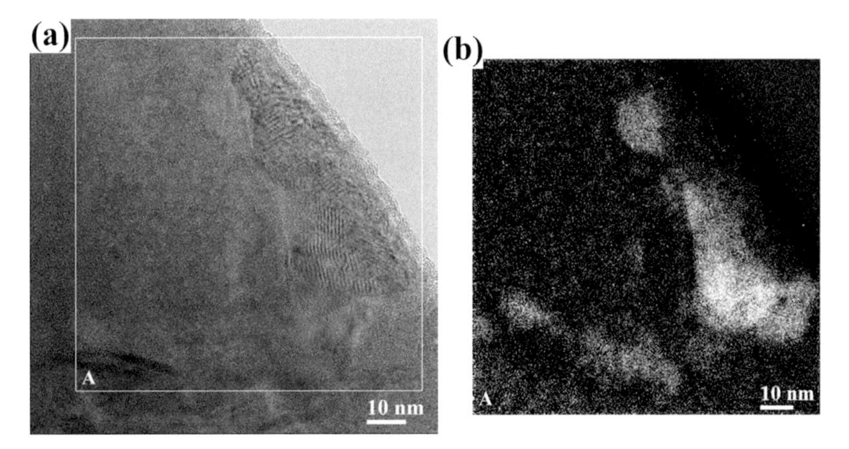

Figure 5.8 (a) HRTEM image of the center of the interface and (b) corresponding EELS Ti map confirming the presence of titanium nanometric grains at the center of the interface.

phases. The general aspect of the interface is also slightly different from that obtained in joints using Ti/Al multilayers and produced under the same conditions. Nevertheless, both interfaces are constituted by the final product of the reaction of the multilayer at the center of the interface and a reaction layer between the multilayers and the γ-TiAl base material.

Based on the results obtained for the joints processed at 800 and 900°C, Ni/Al reactive multilayers having a bilayer thickness of 14 nm appear to be the most suitable for diffusion bonding of γ-TiAl alloy. In fact, these multilayers promoted the formation of a sound joint with an imperceptible bond line. These multilayers were tested at 900°C at a lower pressure (5 MPa) and a shorter bonding time (30 min) and they were shown to be effective in the production of sound joints. The microstructure of the interface obtained under these processing conditions is shown in Figure 5.9.

The interface can be divided into three distinct layers (as presented in Figure 5.9 (b)): a layer of columnar grains close to the base material, a layer of reduced thickness composed of nanometric grains, and the central region composed of equiaxed ultrafine grains. The main difference observed at this interface is the layer with nanometric grains located between the columnar and equiaxed grains. Phase's identification revealed that the columnar layer close to the γ-TiAl is composed of AlNiTi + Al_2NiTi (the same phases identified for a joint produced over 60 min), while an NiAl phase is identified in the other two layers.

The difference in the grain size of the two layers consisting of NiAl can be attributed to the higher titanium amount detected in the layer with nanometric grains. The titanium content in the region adjacent to the columnar layer increases from 2.3 per cent (bonding time of 60 min) to

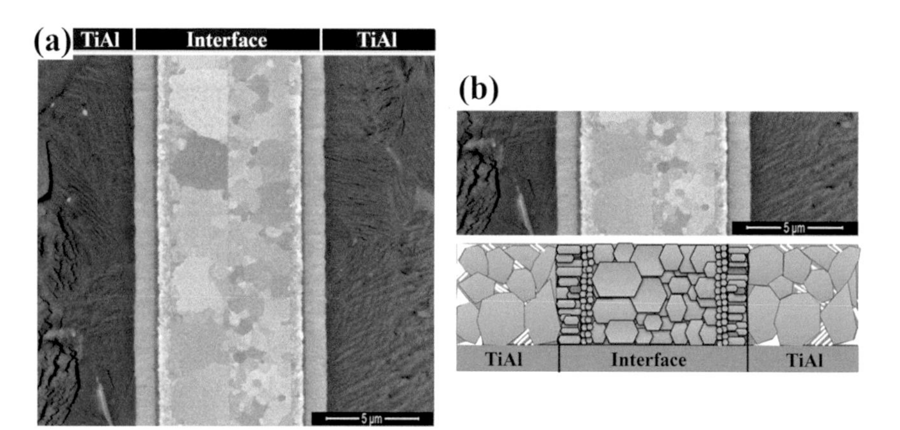

Figure 5.9 (a) SEM image of the bond interface produced by diffusion bonding at 900°C for 30 min and under a pressure of 5 MPa using Ni/Al reactive multilayers with a bilayer thickness of 14 nm and (b) detail and schematic illustration of the layers that constitute the interface.

7.3 per cent (bonding times of 30 min). The titanium amount in this layer appears to inhibit grain growth.

Increasing the bonding temperature to 1,000°C, under 5 MPa for 60 min, promoted the formation of a defect-free interface. The interface was composed of the same layers observed at 900°C. However, the microstructure of the γ-TiAl alloy had been changed due to the deformation caused by the bonding conditions and the intense diffusion with the multilayers. This effect on the base material was not observed in the works of Duarte et al. (2006a,b) and Cao et al. (2008) and can be associated with the use of a γ-TiAl alloys with a lower amount of niobium in both studies.

In short, the microstructural characterisation showed that the multilayers with 14 nm of bilayers thickness are the most promising for producing sound joints under lower bonding conditions. At 800 and 900°C, γ-TiAl alloys can be joined successfully using these Ni/Al reactive multilayers. The required pressure is also very low, 5 MPa at 900°C and 10 MPa at 800°C.

The influence of the bonding conditions was evaluated by nanoindentation and shear tests. The hardness evaluation (Figures 5.10

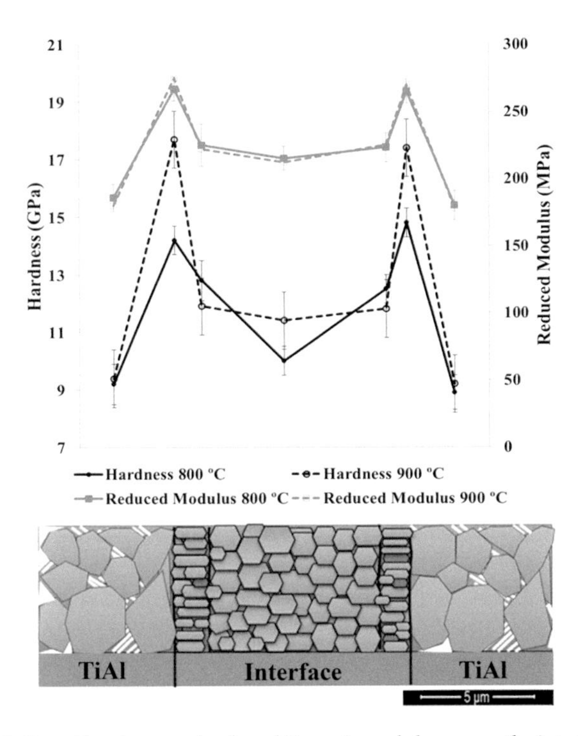

Figure 5.10 Evolution of hardness and reduced Young's modulus across the interfaces produced at 800 and 900°C for 60 min.

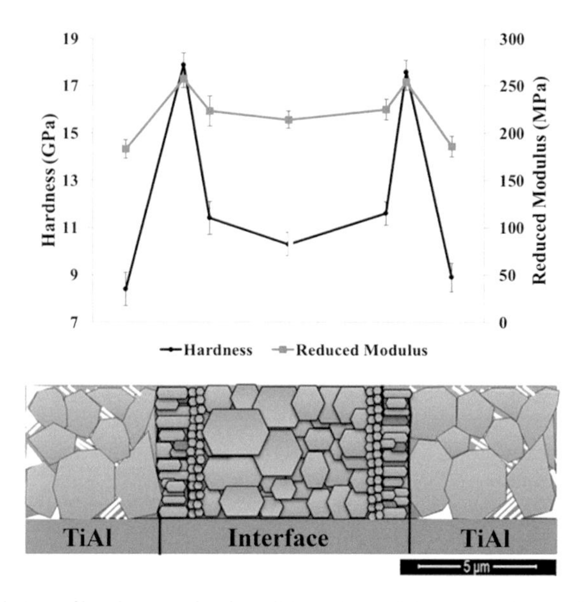

Figure 5.11 Evolution of hardness and reduced Young's modulus across the interfaces produced at 900°C for 30 min.

and 5.11) revealed that the region consisting of NiAl grains had hardness values similar to those of the base material. The layer composed of columnar grains was the hardest zone of the interface. This can be attributed to the Al-Ni-Ti intermetallic compounds formed in this zone. Increasing the bonding temperature from 800 to 900°C led to an increase in the hardness value of this layer from 14 to 18 GPa, respectively. The explanation for this increase is related to the new phase (Al_2NiTi) formed at the higher temperature. The presence of very hard intermetallic compounds can be negative for the mechanical behaviour of the joints. The reduced Young's modulus values confirmed the presence of different phases at the interface. This was expected, since the different phases that composed the interface and the base materials exhibited a different Young's modulus (Kipp 2010). The two bonding times studied (30 and 60 min) at 900°C produced joints with identical hardness values.

The shear strength values of the bonds produced using different processing conditions and multilayers with different bilayer thicknesses are listed in Table 5.1. Some joints fractured during the preparation of the shear test samples and therefore only the promising bonding conditions are shown in this Table.

The joint produced at 800°C for 60 min under a pressure of 10 MPa, with 14 nm of bilayer thickness, showed the maximum shear strength value of 560 MPa. However, these joints revealed the highest dispersion of shear strength values. This dispersion (46 to 560 MPa) was an indication

Table 5.1 Shear strength values obtained for the joints produced under different bonding conditions.

Bonding conditions (Temperature/time/pressure/ multilayer bilayer thickness)	**Shear strength (MPa)**		
800°C/60 min/10 MPa/14 nm	46	560	70
900°C/60 min/5 MPa/14 nm	267	331	342
900°C/60 min/5 MPa/30 nm	46	49	42
900°C/30 min/5 MPa/14 nm	149	201	247

of the unreliability of the bonding, as it had areas well bonded with higher shear strength, but also unbonded regions that result in a lower shear strength value. However, the average shear strength value of 229 MPa was similar to that of 239 MPa reported in the literature for joining γ-TiAl alloys without multilayers at 1,000°C for 60 min under a pressure of 10 MPa (Cao et al. 2008).

The increase in temperature led to the production of reliable joints with an average shear strength value of 314 MPa. This value was similar to that obtained for joints without multilayers processed at 1,000°C for 60 min under a pressure of 20 MPa (333 MPa) (Çam et al. 2006). Decreasing the bonding time to 30 min, with 14 nm of bilayer thickness, caused a decrease in the average shear strength to 199 MPa. At this temperature, and contrary to what would be expected taking into account the microstructural characterisation, the mechanical properties were strongly affected by the multilayers' bilayer thickness. The shear strength value decreased abruptly with the increase in the bilayer thickness from 14 to 30 nm. Based on the results of the mechanical characterisation, the better bonding conditions were 900°C for 60 min under a pressure of 5 MPa, using Ni/Al reactive multilayers with 14 nm of bilayer thickness. In fact, the values achieved for the shear strength were higher than those reported in the literature for diffusion bonding with Ti/Al multilayers (Cao et al. 2008) and similar to that obtained under more demanding conditions without multilayers (Çam et al. 2006).

To better understand the results of the shear strength tests, a 3D fracture surface analysis of these samples was performed. Figure 5.12 shows the fracture surfaces of the samples processed under different bonding conditions (800 and 900°C for 30 and 60 min) and with two different bilayer thicknesses (14 and 30 nm). The roughness of these surfaces varied significantly with the bonding processing conditions and the bilayer thickness of the multilayers. A higher roughness was observed for joints produced at a lower temperature and with 30 nm of bilayer thickness. These results are consistent with the shear strength values, since these fracture surfaces correspond to the lower values. The joints produced with 14 nm

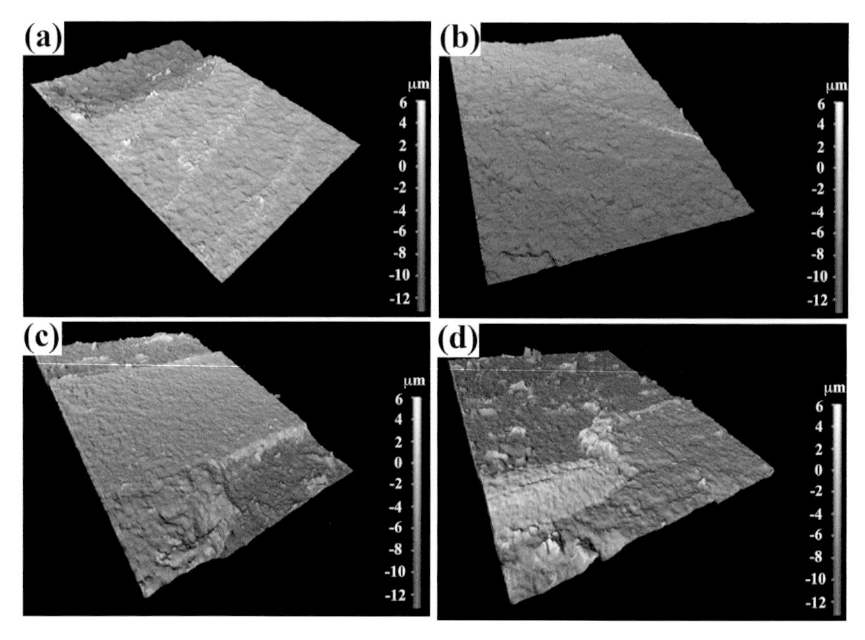

Figure 5.12 3D fracture surface analyses of samples with 14 nm of bilayer thickness bonded at: (a) 800°C for 60 min, (b) 900°C for 60 min and (c) 900°C for 30 min. (d) Sample bonded at 900°C for 60 min with 30 nm of bilayer thickness.

of bilayer thickness at 900°C for 60 min exhibited a very smooth fracture surface. SEM observations showed that the fracture surface is smooth when the crack propagation occurred between the γ-TiAl alloy and the layer composed of Al_2NiTi+AlNiTi columnar grains, while the rougher surfaces are the result of crack propagation across the entire interface.

These results showed that Ni/Al reactive multilayers applied to diffusion bonding of γ-TiAl alloys led to better results than when Ti/Al multilayers were used. The Ni/Al reactive multilayers with a bilayer thickness of 14 nm allowed the reduction in bonding temperature and pressure without impairing the mechanical properties. Although the chemical composition of the multilayers was different from the base material, this did not appear to affect the obtaining of a joint of appreciable mechanical properties. The possibility of further improving the diffusion bonding of γ-TiAl alloys using multilayers can be exploited using a multilayer system with a chemical composition different from that of the materials being joined.

5.4 Diffusion Bonding of γ-TiAl Alloys using Ni/Ti Reactive Multilayers

Ni/Ti reactive multilayers have been the subject of several studies aimed at assessing its use in bonding of similar and dissimilar materials. Although the Ni/Ti reactive multilayers are not a very exothermic system, in fact they have shown promise in reducing the bonding processing conditions (Simões et al. 2013, 2016a,b). Diffusion bonding of γ-TiAl alloy (Ti-45Al-5Nb at.%) with Ni/Ti reactive multilayers has also been investigated. Using Ni/Ti reactive multilayers with 30 and 60 nm of bilayer thickness allows the diffusion bonding of γ-TiAl alloy at 800°C for 60 min under a pressure of 10 MPa. The microstructures of the joint interfaces can be observed in Figure 5.13.

The microstructural analysis showed that interfaces with apparent soundness were produced using both bilayer thicknesses (30 and 60 nm) of Ni/Ti reactive multilayers at 800°C under a pressure of 10 MPa. The microstructure of the interface depended on the bilayer thickness of the multilayer. The joints produced with 30 nm of bilayer thickness had an interface composed of a thicker central region, with a thickness of about

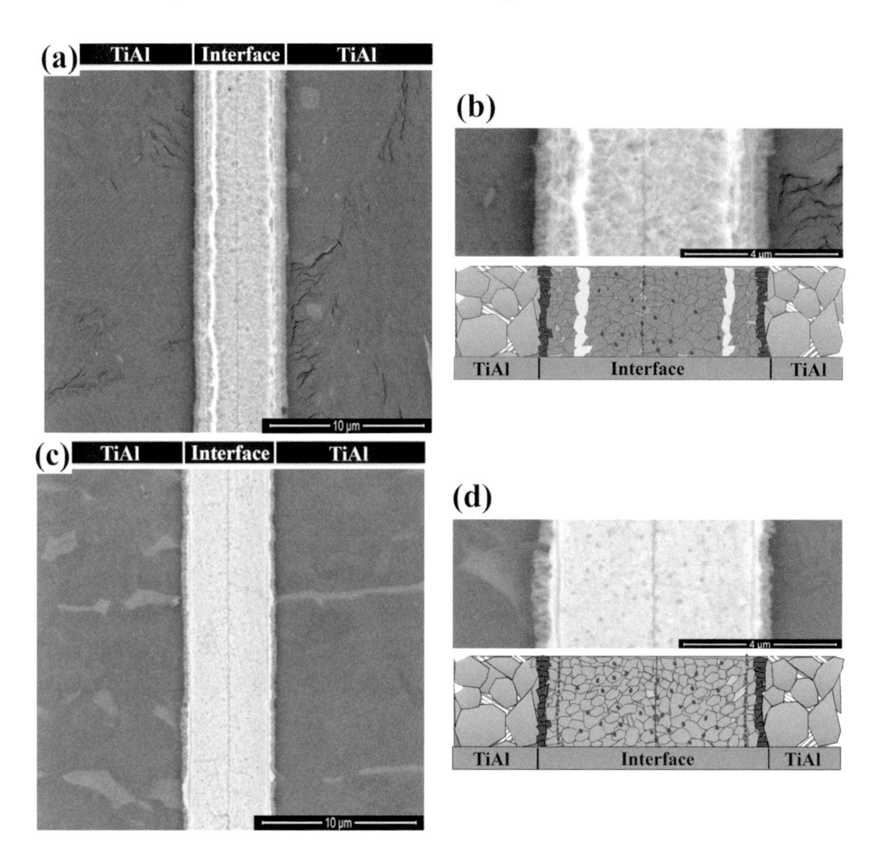

Figure 5.13 (a) and (c) SEM images of the bond interface of the diffusion bonding at 800°C for 60 min under a pressure of 10 MPa using Ni/Ti reactive multilayers with a bilayer thickness of 30 and 60 nm, respectively and (b) and (d) details and schematic illustrations of the layers that constitute the interfaces produced with a bilayer thickness of 30 and 60 nm.

2.5 μm, divided by a very thinner dark line, a bright layer and two thin layers close to the base material, with a global thickness of about 0.6 μm. The interface of bonds produced with multilayers with 60 nm of bilayer thickness is similar, but the bright layers were not observed. In addition, the central region of the interface was brighter than the corresponding region for joints produced with 30 nm of bilayer thickness. The higher brightness observed in the Electron Backscattered (BSE) image indicated that the region has a higher mean atomic number, which suggested a higher nickel concentration.

The results from EDS in the TEM at the interfaces of the joints produced with Ni/Ti reactive multilayers with 30 nm of bilayer thickness infered that the thicker central region of the interface is comprised of a mixture of B2-NiTi and $NiTi_2$ equiaxed grains resulting from the reaction of the Ni/Ti multilayers (Figure 5.14).

Cavaleiro et al. (2014) studied the *in situ* evolution of the Ni/Ti multilayers with temperature and observed the formation of the B2-NiTi and $NiTi_2$ phases, regardless of the bilayer thickness. Aligned nanometric grains of $NiTi_2$ constituted the thin dark line at the center of the interface. This occurs due to the intense diffusion of titanium from the base material. The bright layer has been identified as B2-NiTi and the two layers closest to the base material as $AlNi_2Ti$ and AlNiTi. The formation of these layers reveals the interdiffusion of aluminum, titanium and nickel across the interface.

For the joint produced with a multilayer with 60 nm of bilayer thickness, the phases identified at the interface, are practically the same. The only

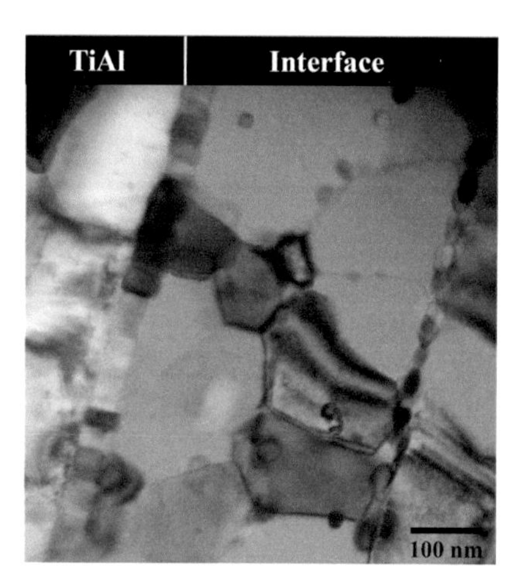

Figure 5.14 TEM image of the bond interface of the diffusion bonding at 800°C for 60 min under a pressure of 10 MPa using Ni/Ti reactive multilayers showing the microstructure of the interface close to the base material, the mixture of B2-NiTi and $NiTi_2$ grains and aligned nanometric grains of $NiTi_2$.

exception was the absence of the bright B2-NiTi layer observed at the interface of the joint produced with 30 nm of bilayer thickness.

With a bonding temperature of 700°C, the bonding was effective only with an increase in pressure to 50 MPa, using the multilayers with 30 nm of bilayer thickness. In Figure 5.15 the microstructure obtained for the joints processed at 700°C for 60 min under a pressure of 50 MPa can be observed. The microstructure of this interface is similar to the one obtained with the temperature of 800°C.

The identification of the phases has shown that the interface produced at 700°C has the same phases in the central zone. However, the two layers closest to the γ-TiAl alloy had a different chemical composition, being constituted of B2-NiTi grains (bright region) and a mixture of B2-NiTi and $NiTi_2$ grains (region adjacent to the γ-TiAl alloy). Low bonding temperature limited the diffusion between the multilayer and the base material, thus inhibiting the formation of the Al-Ni-Ti intermetallic phases.

These interfaces were more complex, displaying more zones than the joints produced with multilayers of Ni/Al and Ti/Al (Duarte et al. 2006a,b, Ustinov et al. 2008, Simões et al. 2010, 2012, 2015). However, the temperature required to produce sound bonds with a Ni/Ti reactive multilayer is lower: 800°C compared with 900°C and 1,000°C for joints produced with Ni/Al and Ti/Al reactive multilayers, respectively.

Mechanical properties were evaluated through nanoindentation and shear tests. The hardness evaluation revealed that the interfaces produced at different bonding temperatures and with different bilayer thicknesses exhibited a similar hardness distribution across the interface (Figures 5.16 to 5.18). The highest hardness value was measured for the γ-TiAl base material. The

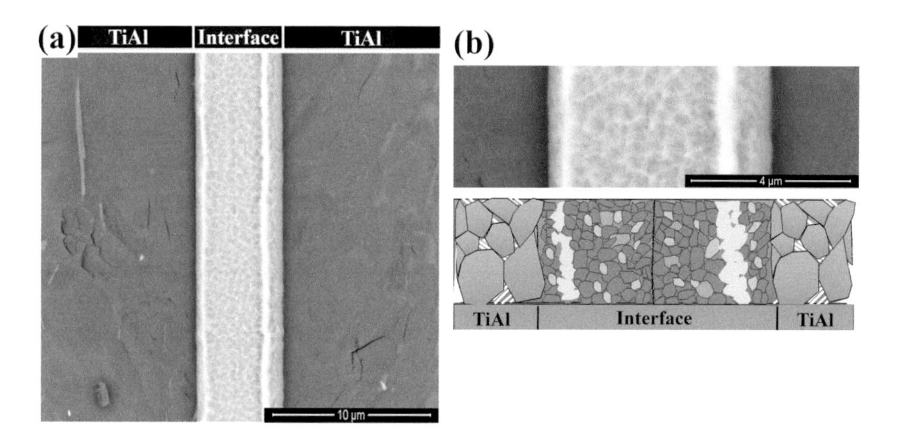

Figure 5.15 (a) SEM image of the bond interface of diffusion bonding at 700°C for 60 min under a pressure of 50 MPa using Ni/Ti reactive multilayers with a bilayer thickness of 30 nm and (b) detail and schematic illustration of the layers that constitute the interface.

region composed of a mixture of B2-NiTi and $NiTi_2$ grains exhibited the lowest hardness values. The reduced Young's modulus distribution across the interface was similar to the distribution of hardness, which is consistent with the presence of different phases across the interface.

The lowest value of reduced Young's modulus observed in the center of the interface was expected, since the different phases that constitute this region have lower Young's modulus than the γ-TiAl base material (Kipp 2010, Toprek et al. 2015).

The mechanical strength of the joints was evaluated through shear tests. Table 5.2 displays the shear strength values of the bonds produced at 800°C using Ni/Ti reactive multilayers with 30 and 60 nm of bilayer thickness. The strength of the joints processed at 700°C was not determined, since the specimens fractured at the beginning of the test, evidencing a discontinuous and weak bonding. The higher shear strength value was observed for the joints produced with multilayers with 30 nm of bilayer thickness (288 MPa). This can be explained by the existence of some unbonded areas at the edge of the samples produced with the bilayer thickness of 60 nm. These areas acted as crack initiation sites, thus degrading the mechanical properties. The best mechanical resistance obtained for the smallest bilayer thickness was the result of thinner individual layers of nickel and titanium, and consequently lower diffusion distances for complete reaction of the multilayer, which

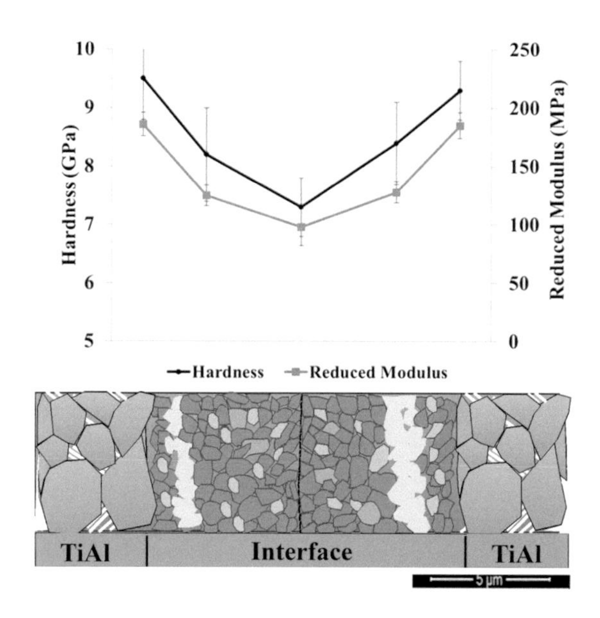

Figure 5.16 Evolution of hardness and reduced Young's modulus across the interfaces produced at 700°C under a pressure of 50 MPa using Ni/Ti reactive multilayers with a bilayer thickness of 30 nm.

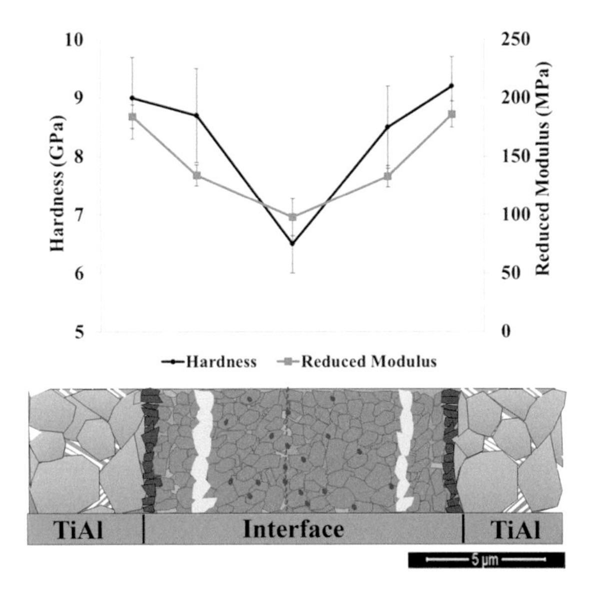

Figure 5.17 Evolution of hardness and reduced Young's modulus across the interfaces produced at 800°C under a pressure of 10 MPa using Ni/Ti reactive multilayers with a bilayer thickness of 30 nm.

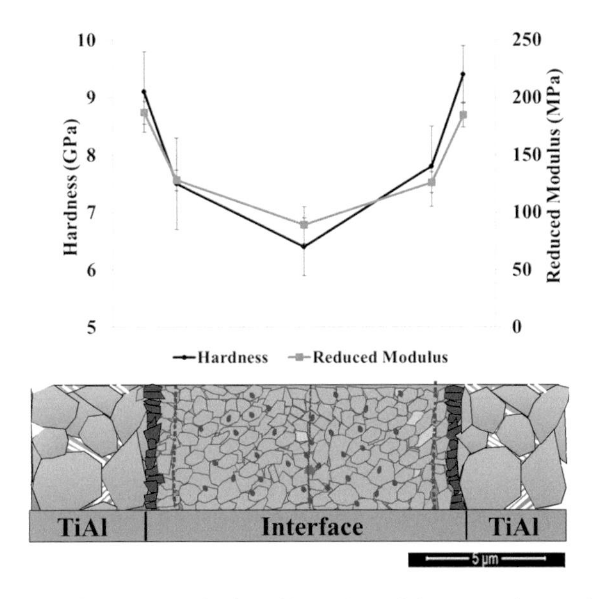

Figure 5.18 Evolution of hardness and reduced Young's modulus across the interfaces produced at 800°C under a pressure of 10 MPa using Ni/Ti reactive multilayers with a bilayer thickness of 60 nm.

Table 5.2 Shear strength values obtained for the joints produced with Ni/Ti reactive multilayers.

Bonding conditions (Temperature/time/pressure/ multilayer bilayer thickness)	**Shear strength (MPa)**		
800°C/60 min/10 MPa/30 nm	258	288	318
800°C/60 min/10 MPa/60 nm	134	140	145

favours the bonding process. The shear strength values clearly shows that the use of Ni/Ti multilayers in the diffusion bonding of γ-TiAl alloys allowed bonding under much less demanding processing conditions. In fact, the shear strength value of 288 MPa achieved with Ni/Ti reactive multilayers at 800°C was only obtained at a temperature of 900°C when Ni/Al reactive multilayers were being used. Without multilayers, even more demanding bonding conditions are needed for a similar mechanical resistance: 950°C for 180 min under 10 MPa or 1,000°C for 60 min under 10 MPa.

To sum up, the diffusion bonding of γ-TiAl can be improved using different reactive multilayers. Despite the use of Ni/Ti reactive multilayers leading to the formation of a more complex interface, these multilayers are those that allow diffusion bonding of γ-TiAl alloys under less severe processing conditions, making this joining process economically attractive.

5.5 Concluding Remarks

This chapter has described a new approach to diffusion bonding processes of γ-TiAl alloys, which consists in the use of reactive multilayers as interlayers to improve diffusion across the interface and produce high strength joints. The reactive multilayers can be deposited on the materials to be joined, to take the advantage of their excellent adhesion to the base material. This approach allows the reduction of the bonding processing conditions due to two characteristics of the multilayer: its reactivity and its nanometric character. In fact, the layers of the two different metallic materials react at low temperature, releasing heat that can be used as a localised heat source, assisting the bonding process. These multilayers, in addition to enhancing contact between the mating surfaces, also increase diffusivity at the interface, due to their nanometric size.

Different reactive multilayers can be used to improve the diffusion bonding of γ-TiAl. Successful joints were obtained using Ti/Al, Ni/Al and Ni/Ti reactive multilayers. During bonding the multilayers evolve to a nanostructured interlayer of intermetallic phases. The bilayer thickness of the multilayer influences the bond quality and strength. The interfaces produced with the multilayers can be very thin.

The advantage of using Ti/Al reactive multilayers in diffusion bonding of γ-TiAl alloys is that they have a composition similar to that of the γ-TiAl alloy, leading to the formation of an interface without chemical discontinuities. The use of Ti/Al reactive multilayers was shown to be effective in allowing the bonding of γ-TiAl alloys without defects at 900 and 1,000°C. However, the improvement in the diffusion bonding process was not very significant and better results were obtained with Ni/Al and Ni/Ti reactive multilayers. This can be explained by the fact that heat released by the Ti/Al multilayer reaction is small when compared to the other reactive multilayer systems investigated. The required pressure is very high, which can have a negative effect when used to bond more ductile γ-TiAl alloys, as is the case with niobium-rich alloys.

Ni/Al reactive multilayers are one of the most reactive systems, and can be successfully applied to diffusion bonding of γ-TiAl alloys. Diffusion bonds can be produced at 800°C under a pressure of 10 MPa and at 900°C under a pressure of 5 MPa. With the use of this system the diffusion bonding pressure and temperature can be reduced. Although sound joints can be produced at 800°C, the large variation in shear strength values reveals the unreliability of the processing conditions. The use of Ni/Al reactive multilayers with 14 nm of bilayer thickness is shown to be effective in the application of diffusion bonding of γ-TiAl at 900°C under a pressure of 5 MPa; the joints revealed a shear strength value (314 MPa) similar to those obtained for diffusion bonding without multilayers at higher temperature and pressure (1,000°C and 20 MPa).

The diffusion bonding temperature can be further decreased using Ni/Ti multilayers. Although, these multilayers are less reactive than Ni/Al multilayers, they promote the formation of joints with mechanical properties similar to the ones obtained with NiAl reactive multilayers (288 MPa) at a lower temperature (800°C). These multilayers improve the diffusivity across the interface, leading to the formation of layers composed of B2-NiTi and $NiTi_2$ nanometric grains.

The key to success, using reactive multilayers in the diffusion bonding of γ-TiAl alloys, is the selection of the appropriate multilayer system and bilayer thickness. For Ti/Al, Ni/Al and Ni/Ti reactive multilayers the most promising bilayer thickness ranges from 4 to 30 nm. The nanometric character and localised heat release by the multilayers reaction can effectively reduce the temperature and pressure of the diffusion bonding process. The following chapters will explore the use of these multilayers in diffusion bonding of γ-TiAl to dissimilar materials, as well as discussing the advantages of using them with brazing alloys.

Keywords: Base materials; bilayer thickness; bonding conditions; diffusion bonding; interface; mechanical properties; microstructure; Ni/Al multilayers; Ni/Ti multilayers; reactive multilayers; Ti/Al multilayers.

5.6 References

Çam, G., G. Ìpekoglu, K.-H. Bohm and M. Koçak. 2006. Investigation into the microstructure and mechanical properties of diffusion bonded TiAl alloys. J. Mater. Sci. 41: 5273–5282.

Cao, J., J.C. Feng and Z.R. Li. 2008. Microstructure and fracture properties of reaction-assisted diffusion bonding of TiAl intermetallic with Al/Ni multilayer foils. Journal of Alloys and Compounds 466: 363–367.

Cavaleiro, A.J., A.S. Ramos, F.M. Braz Fernandes, N. Schell and M.T. Vieira. 2014. *In situ* characterization of NiTi/Ti6Al4V joints during reaction-assisted diffusion bonding using Ni/Ti multilayers. J. Mater. Eng. Perform. 23: 1625–1629.

Duarte, L.I., A.S. Ramos, M.F. Vieira, F. Viana, M.T. Vieira and M. Koçak. 2006a. Solid-state diffusion bonding of gamma-TiAl alloys using Ti/Al thin films as interlayers. Intermetallics 14: 1151–1156.

Duarte, L.I., A.S. Ramos, M.F. Vieira, F. Viana and M.T. Vieira. 2006b. Joining of TiAl using a thin multilayer. Mater. Sci. Forum 514-516: 1323–1327.

Duarte, L.I., F. Viana, M.F. Vieira, A.S. Ramos, M.T. Vieira and U.E. Klotz. 2008. Bonding γ-TiAl alloys using Ti/Al nanolayers doped with Ag. Mater. Sci. Forum 587-588: 488–491.

Duarte, L.I., F. Viana, A.S. Ramos, M.T. Vieira, C. Leinenbach, U.E. Klotz et al. 2012. Diffusion bonding of gamma-TiAl using modified Ti/Al nanolayers. J. Alloy. Compd. 536: S424–S427.

Kipp, D.O. 2010. Metal Material Data Sheets, MatWeb, LLC. http://www.matweb.com/

Ramos, A.S., M.T. Vieira, L.I. Duarte, M.F. Vieira, F. Viana and R. Calinas. 2006. Nanometric multilayers: A new approach for joining TiAl. Intermetallics 14: 1157–1162.

Simões, S., F. Viana, V. Ventzke, M. Koçak, A.S. Ramos, M.T. Vieira et al. 2010. Diffusion bonding of TiAl using Ni/Al multilayers. J. Mater. Sci. 45: 4351–4357.

Simões, S., F. Viana, M. Koçak, A.S. Ramos, M.T. Vieira and M.F. Vieira. 2012. Microstructure of reaction zone formed during diffusion bonding of TiAl with Ni/Al multilayer. J. Mater. Eng. Perform 21: 678–682.

Simões, S., F. Viana, A.S. Ramos, M.T. Vieira and M.F. Vieira. 2013. Reaction zone formed during diffusion bonding of TiNi to Ti6Al4V using Ni/Ti nanolayers. J. Mater. Sci. 48: 7718–7727.

Simões, S., F. Viana, A.S. Ramos, M.T. Vieira and M.F. Vieira. 2015. TEM and HRTEM characterization of TiAl diffusion bonds using Ni/Al multilayers. Microsc. Microanal. 21: 132–139.

Simões, S., F. Viana, A.S. Ramos, M.T. Vieira and M.F. Vieira. 2016a. Microstructural characterization of diffusion bonds assisted by Ni/Ti nanolayers. J. Mater. Eng. Perform. 25: 3245–3251.

Simões, S., A.S. Ramos, F. Viana, M.T. Vieira and M.F. Vieira. 2016b. Joining of TiAl to steel by diffusion bonding with Ni/Ti reactive multilayers. Metals 6: 96-1-11.

Toprek, D., J. Belosevic-Cavor and V. Koteski. 2015. *Ab initio* studies of the structural, elastic, electronic and thermal properties of $NiTi_2$ intermetallic. J. Phys. Chem. Solids 85: 197–205.

Ustinov, A.I., Yu.V. Falchenko, A.Ya. Ishchenko, G.K. Kharchenko, T.V. Melnichenko and A.N. Muraveynik. 2008. Diffusion welding of γ-TiAl based alloys through nano-layered foil of Ti/Al system. Intermetallics 16: 1043–1045.

CHAPTER 6

Joining of γ-TiAl Alloys to Steel

6.1 Introduction

As already mentioned, a limiting factor for the integration of γ-TiAl alloys in advanced aerospace and automotive components is the lack of reliable and efficient joining techniques. The main concern of the automotive industries is the development of novel lightweight structure designs. As long as steel remains the primary material for vehicle structures, bonding between steel and lightweight materials is a key for these new vehicle structure designs. For these applications, the joining of dissimilar structural materials is essential. Despite γ-TiAl alloys being very attractive for these applications, the joining of these alloys to other materials, especially to a structural material such as steel, is very difficult due to their high reactivity and the tendency to form brittle intermetallic phases (Cao et al. 2014).

Only a few techniques can be used to obtain successful joints of γ-TiAl alloys to steel. Fusion welding processes promote the formation of interfaces with brittle intermetallic phases and a tendency for solidification cracking that compromises the mechanical properties of the joint and consequently their performance in service (Ding et al. 2002). Brazing (Noda et al. 1997, Liu and Feng 2002, Li et al. 2005, 2006, He et al. 2006, Dong et al. 2013, 2015a), diffusion bonding (Han and Zhang 2001, He et al. 2002a,b, Morizono et al. 2004, Simões et al. 2016a,b) and most recently friction welding (Lee et al. 2004, Dong et al. 2014, 2015b) are referred to in the literature as the most suitable techniques for dissimilar joining of γ-TiAl alloys. However, as for similar joints, the success of these techniques depends on the use of high temperatures and pressures over a long time.

Besides the fact that the demanding bonding conditions are not attractive for industrial implementation, the dissimilarity in the behaviour of the materials to be joined can easily promote the degradation of the mechanical properties. Therefore, the selection and optimisation of bonding conditions are more complex. There are new approaches to conventional processes, and diffusion bonding assisted by reactive multilayers, is one such option for the production of sound dissimilar bonds with the desired properties. The use of reactive multilayers can reduce the temperature and/or the pressure needed for joining, in a similar way to what happens when γ-TiAl is bonded to itself (see the results presented in Chapter 5).

The reactive multilayers can simultaneously improve diffusivity, due to their nanocrystalline nature and high density of defects, and act as a local heat source, due to the heat released by their exothermic reaction. For dissimilar joints, the multilayers can also act as a diffusion barrier and inhibit or limit the formation of brittle intermetallic compounds.

This chapter will briefly present the joining processes and conditions for obtaining a successful bond between γ-TiAl alloys and steel. The influence of processing conditions on the microstructure and mechanical properties of the joints obtained by different processes will be explored, giving prominence to reaction-assisted diffusion bonding by reactive multilayers. A comparison of this approach with the conventional joining processes will be presented.

6.2 Joining of γ-TiAl Alloys to Steel

The use of fusion welding techniques to join γ-TiAl to steel promotes the formation of weak joints due to the formation of brittle phases, such as TiC and several Ti-Fe intermetallic phases, and high residual stresses due to the martensitic transformation that occurs during cooling from the welding temperature (Xu et al. 1999, Ding et al. 2002, Cao et al. 2014). The key to joining γ-TiAl to steel is to avoid the formation of these brittle phases and relieve the residual stresses at the joints.

Brazing boasts some advantages over fusion welding processes for joining γ-TiAl to steel, as it does not involve the melting of the base materials, the brazing temperature is lower, and residual stresses are accommodated by the filler material that acts as a buffer. The selection of the filler material is fundamental for the success of this process. Different filler materials have been investigated for joining γ-TiAl alloys and steel: Ti-based filler alloys are those most used in similar joining of titanium and its alloys, whereas, in dissimilar joints, the Ag-based alloys produce the best results (Cao et al. 2014). Ag-based alloys exhibit good ductility and play an important role in reducing the residual stresses. However, Ag-based alloys limit the maximum service temperature of the joined components to 400°C, due to the lower creep resistance.

Noda et al. (1997) compared the effects of using a Ti-based alloy (Ti-15Cu-15Ni, wt.%) or an Ag-based alloy (Ag-35.2Cu-1.8Ti, wt.%) in the induction brazing of γ-TiAl alloy (Ti–33.5Al–1.0Nb–0.5Cr–0.5Si, wt.%) to structural steel (Fe–0.4C–0.7Mn–0.2Si–1.8Ni–0.8Cr–0.2Mo, wt.%). In this study, brazing at 15°C above the liquidus temperature of the fillers for 30 s produced sound bonds with both filler materials. At the interface obtained with the Ag-based alloy were detected by two phases (an Ag-rich and an $AlCu_2Ti$ phase), while the one produced with the Ti-based alloy was composed of TiC and two Ti-rich phases.

The highest tensile strength (320 and 310 MPa at room temperature and 500°C, respectively) was observed for the joint produced with the Ag-based alloy. The lower strength of the joints brazed with the Ti-based filler material was associated with the formation of a carbide layer at the interface. Other researchers also obtained unsatisfactory results in the brazing of dissimilar joints of γ-TiAl to steel using Ti-based alloys. Dong et al. (2013) conducted vacuum brazing of γ-TiAl alloys (Ti-48Al-2Cr-2 Nb at.%) to 40Cr steel (Fe-0.4C-1.0Cr-0.7Mn-0.3Si-0.2Ni wt.%) at 900°C for 5, 10 and 15 min with a Ti-based filler alloy (Ti-22Ni-10Cu-8Zr at.%). Distinct zones consisting of α_2-Ti_3Al, Ti_2Ni, γ-TiAl and Ti-rich phases constituted the interface. The shear strength of these joints was very low (26 and 32 MPa for joints produced with dwell times of 5 and 15 min, respectively). The low strength of the joints was attributed to an insufficient interdiffusion between the steel and the filler material, which hindered the bonding. Therefore, a proper selection of the filler material and the processing conditions is necessary to ensure interdiffusion and allow the production of high-strength joints.

Although brazing γ-TiAl and steel with Ag-based fillers exhibited the best results, slight changes in the composition of these fillers had a significant effect on the microstructure as well as on the mechanical properties of the joints. For instance, Liu and Feng (2002) used an Ag-based filler with zinc (Ag-34Cu-16Zn, wt.%) for brazing a γ-TiAl alloy (Ti-43Al-1.7Cr-1.7Nb at.%) to steel (Fe-0.4C-1.0Cr-0.7Mn-0.3Si-0.2Ni, wt.%). This filler material was used to improve wetting and spreading of the filler on the γ-TiAl alloy. The brazing experiments were performed at 900°C for 2 to 40 min in a vacuum. The interface was characterised by $Ti(Cu,Al)_2$ and Ag-based solid solution close to γ-TiAl and Ag-Cu eutectic and TiC close to steel, most of the zinc of the filler was eliminated during brazing, probably by evaporation. The maximum shear strength value (190 MPa) was obtained for joints processed for 20 min. Increase in the brazing time led to a decrease in the shear strength, as the result of the increase in the thickness of the TiC layer.

Li et al. (Li et al. 2005, 2006) also investigated the use of an Ag-based filler alloyed with Ti (Ag-33-Cu-4.5-Ti wt.%) for brazing γ-TiAl (Ti–46.5Al–2.5V–1.0Cr at.%) and 42CrMo steel (Fe–0.42C–1.0Cr–0.7Mn–0.3Si–0.5Mo wt.%). Different layers of $AlCu_2Ti$, AlCuTi and α_2-Ti_3Al intermetallics, Ag-based solid solutions and TiC carbide comprised the joint interface. The highest tensile strength (347 MPa) was observed for joints processed at 900°C for 5 min in a vacuum.

He et al. (2006) investigated the application of the Ag-35.2Cu-1.8Ti wt.% in the brazing of the γ-TiAl (Ti–48Al–2Cr–2Nb and Ti–47.5Al–2.5V–1Cr at.%) to 35CrMo steel (Fe–0.4C–0.3Si–0.6Mn–0.1Cr–0.2Mo wt.%). The brazing experiments were performed at temperatures ranging from 850 to 970°C with dwell times extending from 1 to 10 min. The interfaces presented the typical layered microstructure formed by intermetallic compounds of the Al-Cu-Ti system, with thicknesses strongly influenced by the brazing conditions. The tensile strength decreased with increasing brazing temperature, due to the increase in the thickness of the intermetallic compounds at the interface. According to the authors, the thickness of the $AlCu_2Ti$ layer must be one-third of the thickness of the total interface in order to achieve high strength. The joints produced at 870°C for a dwell time of 5 min exhibited the highest tensile strength (320 MPa).

Recently, Dong et al. (2015a) have investigated the brazing of γ-TiAl (Ti–48Al–2Cr–2Nb at.%) to 40Cr steel (Fe–0.4C–1.0Cr–0.7Mn–0.3Si–0.2Ni wt.%) using newly developed amorphous foils with two compositions: Cu-37.5Ti-6.25Ni-6.25Zr-6.25V and Cu-25Ti-12.5Ni-12.5Zr-12.5V at.% at 900 and 930°C. These new filler materials were developed to: (1) reduce the titanium content of the interface, thus diminishing the possibility of TiC formation, (2) add elements that decrease the probability of formation of Ti-Fe intermetallics, and (3) improve the mechanical properties of the joints at elevated temperatures. The interface obtained with the high titanium content foil, Cu-37.5Ti-6.25Ni-6.25Zr-6.25V, was flat and thin, containing $Ti_{19}Al_6$ and Ti_2Cu phases, while the interface obtained with the other filler foil was uneven, with bulges composed mainly of a Ti-rich solid solution. The higher shear strength observed was 107 MPa for the joints produced with the filler alloy of low titanium content, Cu-25Ti-12.5Ni-12.5Zr-12.5V, due to the absence of intermetallic compounds at the interface.

The results presented in the literature show that the brazing of γ-TiAl to steel is possible using Ag or Ti-based filler alloys. The highest strength joints, 320 to 347 MPa, are produced with Ag-based fillers alloyed with small amounts of titanium, the active metal brazing alloys. Brazing conditions must prevent the formation of brittle intermetallic phases and TiC to achieve the best results. Despite brazing being a method, which produces sound joints between γ-TiAl and steel, the interfaces are always thick, with a non-uniform microstructure and chemical composition, some residual thermal stresses and limitations for high temperature applications.

Diffusion bonding is a promising method for improving the strength of dissimilar joints of γ-TiAl and steel. High bonding temperatures improve the mechanical properties of the joints, since they promote the formation of defect-free interfaces due to the enhancement of diffusion. However, these high temperatures are unsuitable for bonding steel, since they can cause plastic deformation and promote the formation of brittle intermetallic phases that will impair the strength of the joints. Since the base materials exhibit different mechanical behaviours, the process conditions have to be selected carefully, in order to obtain the best compromise between the microstructure and the mechanical properties of the joints.

Han and Zhang (2001) investigated the influence of the bonding conditions (temperature, pressure and time) in the diffusion bonding of a γ-TiAl alloy (Ti-33Al-2Cr-2Nb, wt.%) to 40Cr steel (Fe–0.4C–1.0Cr–0.7Mn–0.3Si–0.2Ni, wt.%). High strength joints were produced at 950 to 1,100ºC for 10 to 30 min under pressures of 5 to 10 MPa. As expected, the increase in the bonding temperature promoted the formation of joints with higher tensile strength (300 MPa at 1,100ºC with a pressure of 10 MPa). Pressure also plays an important role in the quality of the joint, since this is essential for reducing porosity and promoting contact between the mating surfaces, attention being given to the fact that high pressures also cause the formation of cracks in the γ-TiAl. The results demonstrated that the bonding conditions should be optimised to promote diffusion across the interface and produce defect-free joints, avoiding changes in the microstructure of the base materials that compromise the strength of the joints. The selection of the bonding time depends on the temperature and pressure, although it should be sufficient to promote diffusion across the interface. Nevertheless, the bonding conditions depend on the steel chemical composition and whether interlayers are used or not in the diffusion bonding process.

He et al. (2002) also studied the influence of the bonding conditions in the production of defect-free interfaces by diffusion bonding γ-TiAl (Ti–48Al–2Cr–2Nb at.%) to steel (Fe-0.4C-0.3Si-0.7Mn-1.0Cr-0.2Ni at.%). Bonding at temperatures from 850 to 1,100ºC for 1 to 60 min and at pressures from 5 to 40 MPa produced interfaces with a layered microstructure composed of TiC and Ti-Fe-Al intermetallic compounds. The number and thickness of the reaction layers depended on the bonding conditions. The mechanical properties of the joints were closely related to the microstructure of the interface. For joints processed at temperatures higher than 950ºC for dwell times longer than 10 min, the thickness of the TiC and Ti-Fe-Al intermetallic compounds increased quickly and undermined the mechanical strength of the joints. Diffusion bonding at 850ºC exhibited very low tensile strength (18 MPa); increasing the bonding temperature to 950°C for 60 min under a pressure of 20 MPa led to an increase in tensile strength to 100 MPa. This improvement resulted from a more homogeneous chemical composition of the interface. The higher tensile strength values (170–185 MPa) were obtained for the joints processed at 930–960ºC, for 5 or 6 min under pressures of 20 to 25 MPa.

Interfacial microstructures of the diffusion bonds of γ-TiAl (Ti-36Al, wt.%) and eutectoid steel (Fe-0.8C-0.2Si-0.4Mn-0.1Cr, wt.%) were investigated by Morizono et al. (2004). The joining experiments were performed at temperatures from 800 to 1,000ºC for 60 min in a vacuum. The interface comprised four layers: a ferrite layer adjacent to the steel, followed by a TiC layer, and two layers composed of Ti-Al-Fe compounds (Fe_2Al_5 and Fe_3Al) which formed close to γ-TiAl. The highest shear strength value (160 MPa) was observed for the joint processed at 800°C. The increase in the bonding temperatures led to a decrease in the shear strength due to the increase in the thickness of the reaction layers composed of Ti-Al-Fe intermetallic compounds.

Diffusion bonding of γ-TiAl (Ti-45Al-5Nb at.%) to AISI 310 (Fe-0.3C-25Cr-20Ni wt.%) stainless steel has also been successfully produced at 800°C for 60 min, under pressures of 25 and 50 MPa in a vacuum (Simões et al. 2016a). These bonding conditions enabled the production of sound interfaces without affecting the microstructure of the base materials, although unbonded regions were observed at the edge of the joints.

Figure 6.1 shows the Scanning Electron Microscopy (SEM) images of the interface. Four layers were identified: one with columnar grains adjacent to steel and three thin layers adjacent to γ-TiAl. A detail of the interface and a scheme of the four zones can be observed in Figure 6.1 (b). The phase identification performed by High Resolution Transmission Electron Microscopy (HRTEM) and Fast Fourier Transform (FFT) analysis revealed that the region closest to the steel was composed of large grains of α-Fe and a few nanometric NiAl grains (Figure 6.2). These grains were formed by the reaction of aluminum, diffusing from the γ-TiAl base alloy, and nickel diffusing from the steel.

At the center of the interface, a bright layer composed of NiAl grains with α-Fe nanoparticles was observed. The diffusion of aluminum and titanium from γ-TiAl and nickel from steel resulted in the formation of thin layers of $AlNi_2Ti$ and Al_2NiTi close to the γ-TiAl base material. The formation of FeTi intermetallics or TiC was not detected. The interface was harder than the base materials, as can be confirmed by observing Figure 6.3.

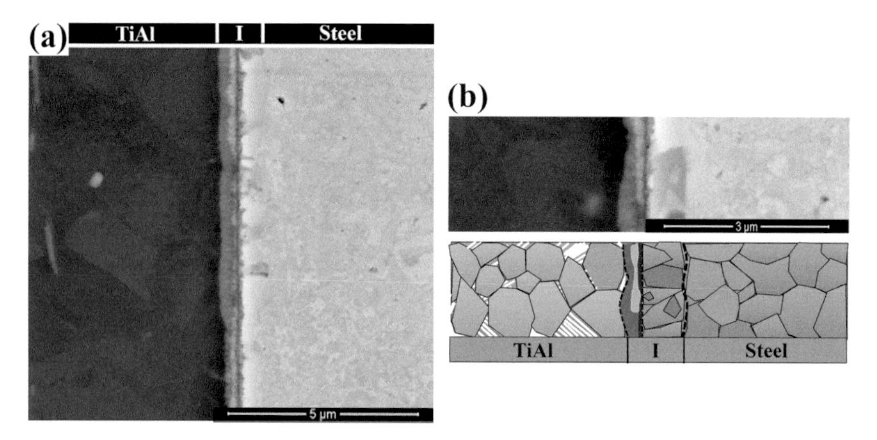

Figure 6.1 SEM images of the bond interface of joints between γ-TiAl and steel obtained at 800°C, with 25 MPa for 60 min: (a) low magnification of the joint and (b) a detail of the interface (I) and a scheme of the four zones observed.

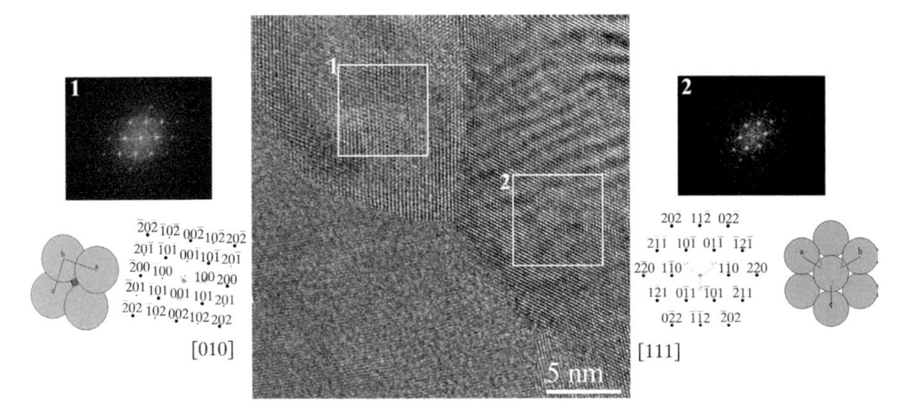

Figure 6.2 HRTEM image and FFT analysis results of two grains at the region closest to steel. Grain 1 is indexed as NiAl and the grain 2 as α-Fe.

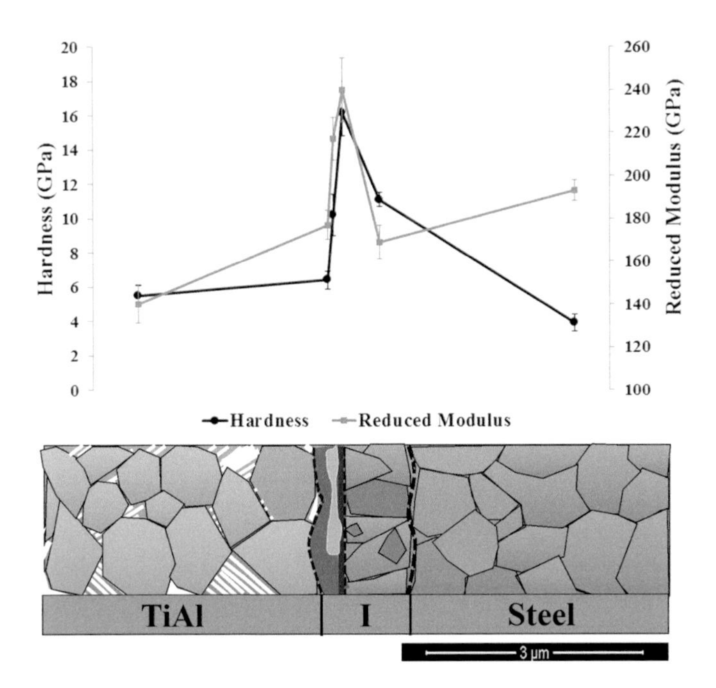

Figure 6.3 Hardness and reduced Young's modulus distributions across the diffusion bond processed at 800°C, with 25 MPa for 60 min.

The highest hardness values (16 GPa) of the interface were measured in the layer composed of small grains of NiAl and dispersed α-Fe nanoparticles. The values of the reduced Young's modulus of each layer corresponded with those reported in the literature for the phases that constitute those layers (Kipp 2010). The low shear strength, close to 37 MPa, of the joints produced at 800ºC for 60 min and at a pressure of 25 MPa was attributed to the unbonded regions and the formation of harder and more brittle intermetallic phases at the interface. This shear strength value was similar to those reported by other authors for γ-TiAl to steel joints produced by brazing at 900°C and diffusion bonding at similar or higher temperatures.

The use of interlayers in the diffusion bonding of γ-TiAl to steel can prevent the formation of brittle intermetallic compounds, acting as a diffusion barrier, and improve contact between the surfaces, leading to a more complete joint between the base materials. He et al. (2002b, 2003) investigated the influence of a composite barrier consisting of a stack of foils, vanadium/copper or titanium/vanadium/copper, as an interlayer in the diffusion bonding of γ-TiAl (Ti–48Al–2Cr–2Nb at.%) to 40Cr steel (Fe-0.4C-0.3Si-0.7Mn-1.0Cr-0.2Ni at.%). The joining experiments were performed at temperatures from 800 to 1,100ºC with bonding times ranging from 5 to 60 min and pressures ranging from 5 to 30 MPa. The use of these interlayers allowed the production of sound interfaces at 1,000ºC with 20 MPa for 20 min. Iron and carbon did not diffuse through the vanadium or copper foils, avoiding the formation of TiC and Al-Fe-Ti intermetallic phases, whereas α_2-Ti_3Al and V_2Al_8 formed at the TiAl/V interface. Through the use of the titanium/vanadium/copper interlayer, the formation of V_2Al_8 was inhibited and the interfaces were only composed of α_2-Ti_3Al, γ-TiAl and titanium solid solution. The improvement in the diffusion bonding by the use of interlayers was confirmed by the values of the tensile strength of these joints. The highest value (420 MPa) was obtained for joints processed with the titanium/vanadium/copper interlayer. However, the demanding bonding conditions remain unattractive.

Friction welding is another process that enables the successful production of dissimilar joints between γ-TiAl and steel. This process has considerable advantages, such as high quality, precision, efficacy and energy saving. However, a careful selection of the friction parameters is crucial for all materials. Dong et al. (2014, 2015b) produced joints of γ-TiAl (Ti–48Al–2Cr–2Nb at.%) to 40Cr steel (Fe–0.4C–1.0Cr–0.7Mn–0.3Si–0.2Ni wt.%) by friction welding at a constant rotational speed of 1,500 rpm. Sound joints were obtained, but their strength was very low (86 MPa) owing to the formation of the TiC and martensite. Post-Bond Heat Treatment (PBHT) at 580ºC for 2 hr increased the strength to 395 MPa. The use of filler materials as interlayers can also be used to enhance the mechanical properties of the friction welding joints, as demonstrated by the work of Lee et al. (2004). The aim of the use of this intermediate material is to avoid the formation of brittle phases and prevent crack formation. Lee et al. (2004) bonded γ-TiAl (Ti–47Al at.%) and AISI 4140 steel (Fe-0.4C-0.3Si-0.7Mn-1.0Cr-0.2Ni at.%) by friction welding with and without pure copper as an interlayer. Due to the martensitic transformation, brittle reaction products and residual stresses, resulted in cracking through the interface of the bonds produced without copper. The use of the copper interlayer avoided the martensitic transformation and therefore cracks formation. Moreover, the heat-affected zone in the steel was minimised by the use of the copper interlayer. These joints presented a tensile strength of 375 MPa. Although this process does not involve the melting of the base material, the microstructural changes are very significant and closely depend on the processing conditions and characteristics of the base materials.

A summary of the joining processes and bonding conditions for dissimilar bonding of γ-TiAl and steel, as well as the mechanical properties of the joints, can be seen in Table 6.1.

Diffusion bonding is the process that produces joints with the highest mechanical properties. However, the demanding bonding conditions, which are necessary, makes the processes less attractive. The production of joints with similar mechanical properties can also be obtained by brazing and friction welding; however, a careful selection of the filler material for brazing, and the need for a PBHT or the use of interlayers for the friction processes, are mandatory requirements for obtaining high mechanical properties.

One possibility for obtaining joints with higher mechanical properties under less stringent conditions is diffusion bonding assisted by reactive multilayers. In the next section, the diffusion bonding of γ-TiAl alloys to steel with multilayers will be explored in detail.

6.3 Diffusion Bonding of γ-TiAl Alloys to Steel with Reactive Multilayers

The use of base materials coated with reactive multilayers is an alternative to reducing the temperature, time and/or the pressure needed in the diffusion bonding process, as already evidenced in the joining of γ-TiAl alloys.

As described in Chapter 5, less stringent bonding conditions are the result of two effects of reactive multilayers; they improve the diffusivity, due to their nanocrystalline character and high density of defects, and act as a local heat source, as a result of the heat released by the exothermic reaction of the multilayers. The possibility of using less demanding conditions in dissimilar joining between the γ-TiAl alloy and steel is of utmost importance, since the joining process may cause plastic deformation of the base materials, due to the different mechanical responses of both materials, which may hinder the bond between components.

Different reactive multilayers have been tested in order to achieve sound joints between γ-TiAl alloys and steel (Simões et al. 2016a,b).

An interlayer formed of Ni/Al reactive multilayers was tested in diffusion bonding of γ-TiAl (Ti-45Al-5Nb at.%) to AISI310 (Fe-0.3C-25Cr-

Table 6.1 Joining technology and processing conditions for bonding between γ-TiAl and steel.

Joining technology		Processing conditions					
Process	**Filler material/Interlayers**	**Temperature (°C)**	**Time (min)**	**Pressure (MPa)**	**Shear Strength (MPa)**	**Tensile strength (MPa)**	
Brazing	Ag-35.2Cu-1.8Ti (wt.%)	830	—	—	—	320	Noda et al. 1997
	Ti-15Cu-15Ni (wt.%)	975	—	—	—	210	
Brazing	Ti-22Ni-10Cu-8Zr (at%)	900	5	—	26	—	Dong et al. 2013
			10	—	32	—	
			15	—	32	—	
Brazing	Ag-34Cu-16Zn (wt.%)	900	10	—	100	—	Liu and Feng 2002
			20	—	190	—	
			40	—	150	—	
Brazing	Ag33Cu4.5Ti (wt.%)	900	5	—	—	347	Li et al. 2005, 2006
Brazing	Ag-35.2Cu-1.8Ti (wt.%)	850	5	—	—	150	He et al. 2006
		870		—	—	340	
		950		—	—	50	
Brazing	Cu37.5Ti6.25Ni6.25Zr6.25V (at.%)	900	10	—	38	—	Dong et al. 2015a
	Cu25Ti12.5Ni-12.5Zr12.5V (at.%)	930		—	107	—	
Diffusion Bonding	—	950	15	10	—	275	Han and Zhang 2001
		1,000			—	290	
		1,100			—	300	

Diffusion Bonding	—	850	60	20	—	18	He et al. 2002
		950			—	100	
		960			—	185	
Diffusion Bonding	—	800	60	—	160	—	Morizono et al. 2004
		900			115	—	
		1,000			50	—	
Diffusion Bonding	—	800	60	25	35		Simões et al. 2016a
Diffusion Bonding	V/Cu	1,000	20	20	—	210	He et al. 2003a
Diffusion Bonding	Ti/V/Cu	1,000	20	20	—	420	He et al. 2003b
		1,100			—	150	
Friction welding	—	—	—	—	—	86	Dong et al. 2014
Friction welding	Cu	—	—	—	—	375	Lee et al. 2004

20Ni, wt.%). Figure 6.4 shows the microstructure of this dissimilar joint performed at 800°C for 60 min and 25 MPa under vacuum when using Ni/Al reactive multilayers with 14 nm of bilayer thickness. Under these bonding conditions, the process does not significantly affect the microstructures of the base materials, as can be observed in the SEM images of the interface.

The interface is characterised by four distinct layers: one layer adjacent to the steel, one thicker layer in the center and two thin layers close to the γ-TiAl base material. A schematic representation of the morphology of the zones that compose the interface can be observed in Figure 6.4 (c). The number of reaction zones is equal to that observed in joints with the same bonding conditions, but without the use of multilayers (Figure 6.1). However, the thicknesses and morphology of the zones are quite different. The presence of multilayers improved contact of the mating surfaces, resulting in fewer unbonded regions.

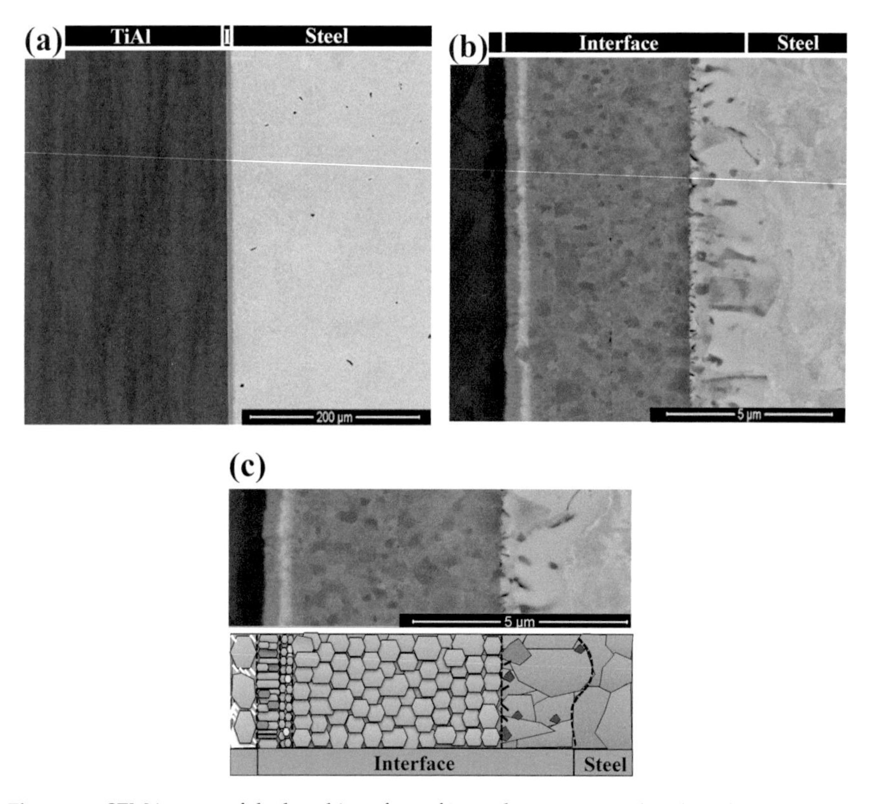

Figure 6.4 SEM images of the bond interface of joints between γ-TiAl and steel using Ni/Al reactive multilayers processed at 800°C with 25 MPa for 60 min: (a) low magnification showing the well-bonded interface, (b) high magnification showing the microstructure of the interface and (c) detail and a scheme of the different zones at the interface.

The identification of phases was performed by Electron Backscatter Diffraction (EBSD) and HRTEM combined with FFT, since some zones are only a few micrometers thick and are formed by phases with submicrometer or nanometer size grains. This identification revealed that the layer adjacent to steel is formed by larger grains of the α-Fe and smaller grains of NiAl. The same phases were observed in the joint produced under identical processing conditions, but without multilayers. This result shows that the use of multilayers did not affect the formation of phases in the layer adjacent to the steel.

NiAl equiaxed grains constitute the thicker layer of the interface. These grains are the result of the reaction between nickel and aluminum multilayers. In similar γ-TiAl joints produced with Ni/Al reactive multilayers, an identical thick layer at the center of the interface was observed. Nanoparticles of chromium and iron were detected in these NiAl grains. This occurred due to diffusion of iron and chromium from the steel to the central zone and subsequent precipitation inside the NiAl grains due to the decrease in solubility with decreasing temperature.

Aluminum, nickel and titanium are the main elements comprising the thin layers close to the γ-TiAl base material. The diffusion of titanium and aluminum from the γ-TiAl alloy promoted the formation of a layer of $AlNi_2Ti$ (bright layer) and another of $AlNiTi + Al_2NiTi$ (dark layer). Figure 6.5 shows an example of phase identification by EBSD analysis of the layers formed close to the γ-TiAl base material.

The influence of the different phases comprising the interface on the mechanical behaviour was evidenced by the hardness and reduced Young's modulus distributions across the interface, as presented in Figure 6.6. For the formation of a joint with high strength, an interface composed of phases with similar properties to those of the base materials is essential. The results given in Figure 6.6 show that the interface layers are slightly harder than the base materials. The values of the reduced Young's modulus of the layers are in accordance with those reported in the literature (Kipp 2010). The use of multilayers led to the formation of an interface with a more uniform mechanical behaviour than that without multilayers.

The shear strength of the γ-TiAl to steel joint produced using the Ni/Al multilayer attained 162 MPa. This value is higher than those reported in relation to the brazing of γ-TiAl to steel (Table 6.1) and to the ones obtained for joints processed by diffusion bonding at the same temperature, but without reactive multilayers (37 MPa). These results can be attributed to an improvement in contact between the mating surfaces, to the formation of different phases at the interface (less brittle phases) and to the absence of Ti-Fe and TiC phases.

Observation of the fracture surfaces of the shear test sample showed that the fracture was initiated at the interface and propagated mainly through the γ-TiAl base material, which confirms good joining between the materials.

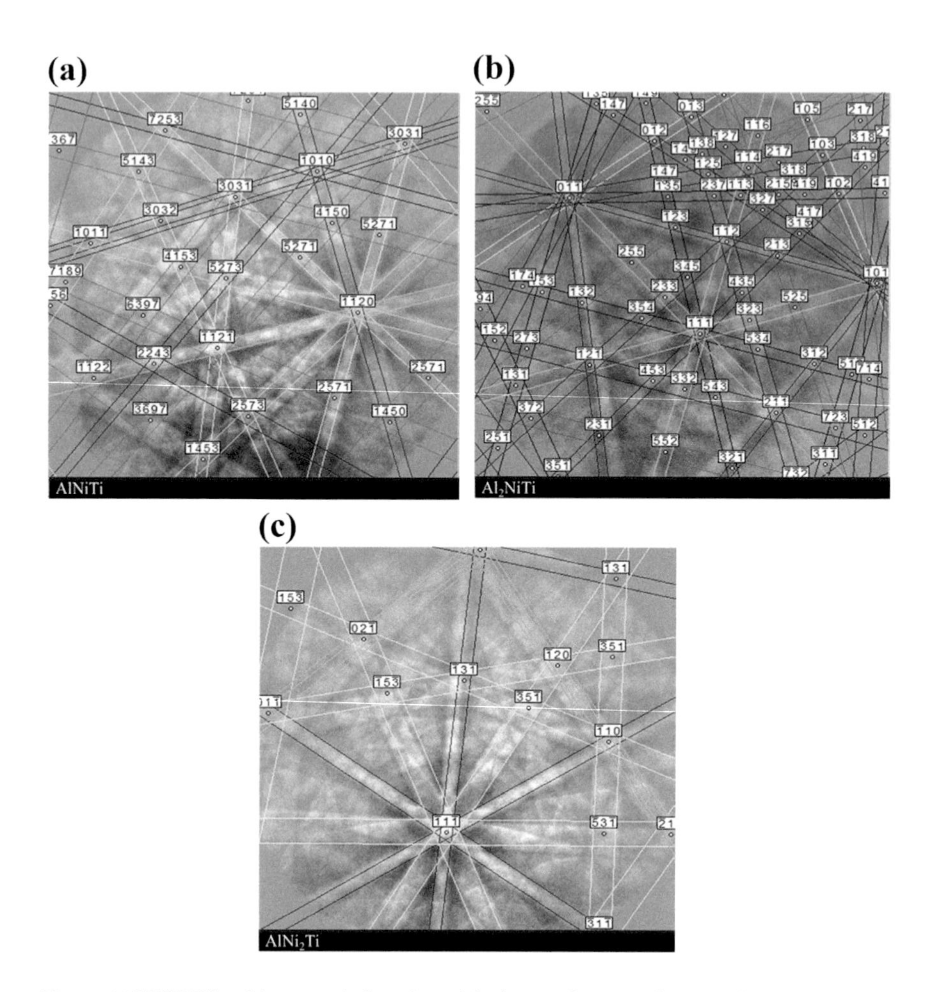

Figure 6.5 EBSD Kikuchi pattern indexation of the layers closest to the γ-TiAl base material of a joint processed at 800°C using Ni/Al reactive multilayers of (a) and (b) the dark layer and (c) the bright layer observed in Figure 6.4.

The use of the Ni/Al reactive multilayers was effective only when the multilayers coated both base materials.

Summing up, diffusion bonding of γ-TiAl to steel using Ni/Al reactive multilayers can be produced under less demanding processing conditions and with higher shear strength values than those produced in the same conditions without multilayers. Despite this improvement in the joining process, the use of the multilayers does not completely eliminate all unbonded areas. These areas reduce the actual bonding area and can act as stress concentrators, which promote failure and impair the shear strength. This process may be more effective if contact between the surfaces to be

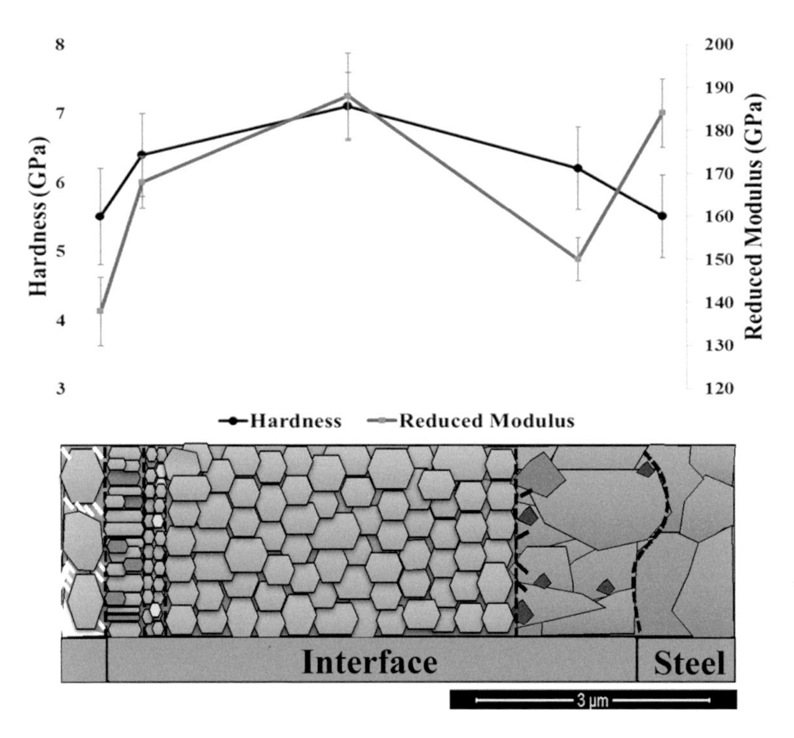

Figure 6.6 Hardness and reduced Young's modulus distributions across the TiAl/steel diffusion bond interface processed at 800°C with 25 MPa for 60 min.

joined is improved, which can be achieved by using a different multilayer system.

Ni/Ti reactive multilayers can be employed as an interlayer to promote the diffusion bonding of γ-TiAl (Ti-45Al-5Nb at.%) to AISI 310 (Fe-0.3C-25Cr-20Ni wt.%), thus eliminating the unbonded areas while using undemanding bonding conditions. SEM images of Figure 6.7 show the microstructure of interfaces of joints produced by diffusion bonding of γ-TiAl alloys to stainless steel using these Ni/Ti reactive multilayers. The joints were processed at 700 and 800°C for 60 min.

The microstructure of the interface is similar for the bonds produced at the two temperatures. The main difference is the thickness of the layers that comprise the interface. This interface reveals more layers than those produced using Ni/Al multilayers. In fact, it consists of six layers: two thin layers close to each of the base materials and a thicker layer at the center of the interface, divided by a thin dark line.

The identification of the phases revealed that the interface is mainly composed of B2-NiTi and $NiTi_2$. Different phases were observed in the thinner layers close to the base materials. AlNiTi and $AlNi_2Ti$ were identified

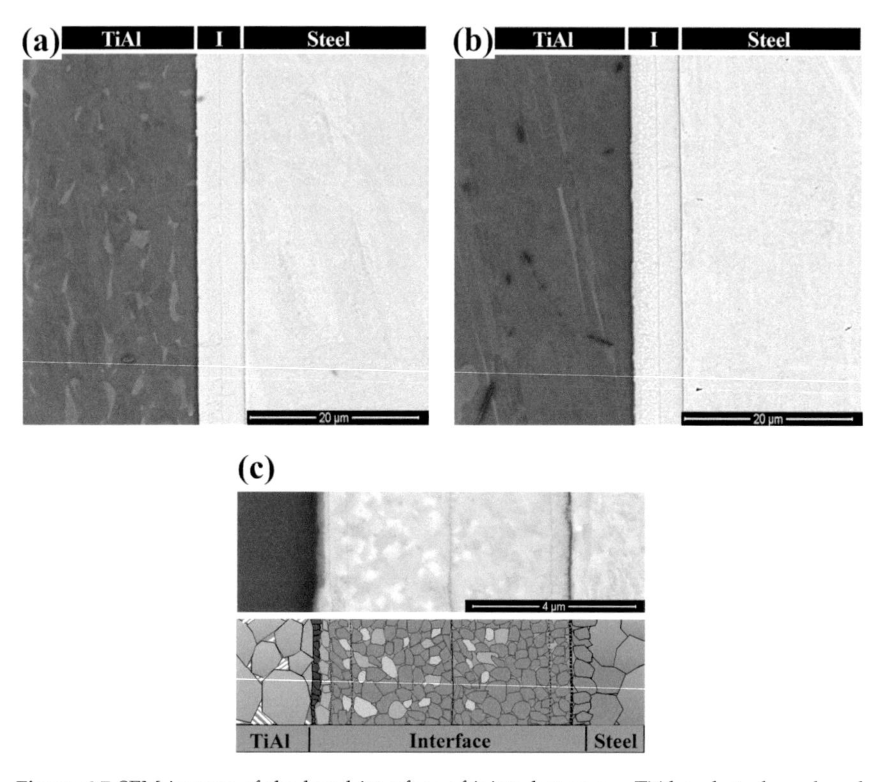

Figure 6.7 SEM images of the bond interface of joints between γ-TiAl and steel produced with Ni/Ti multilayers at (a) 700ºC, (b) 800ºC and (c) detail and scheme of the different zones observed at the interface.

nearest to the γ-TiAl alloy. The formation of layers with these phases showed the interdiffusion of aluminum, titanium and nickel across the interface. Close to the steel, α-Fe and $NiTi_2$ were identified. Some of these phases were also observed in the joints produced using Ni/Al multilayers, such as α-Fe and AlNiTi. However, the use of a different multilayer led to the formation of distinct phases, primarily at the central region of the interface (composed of B2-NiTi + $NiTi_2$). Another distinct feature that should be noted is the identification of nanometric $FeTi_2$ grains at the dark layer closest to the steel. This phase, undetected in joints produced using Ni/Al reactive multilayers, can be explained by the presence of titanium in the Ni/Ti reactive multilayers. During diffusion bonding, titanium combines with iron, diffusing from the steel, and forms this intermetallic.

Microstructural characterisation indicates that the use of Ni/Ti reactive multilayers enables the production of sound joints between γ-TiAl and stainless steel at a lower temperature (700ºC instead of 800ºC for Ni/Al) or under lower pressure at the same temperature (at 800°C, a pressure of 10 MPa is required compared with 50 MPa for joints produced with Ni/Al).

The mechanical properties of the joints produced using Ni/Ti reactive multilayers were also evaluated by nanoindentation and shear tests. Figure 6.8 shows the distribution of hardness and reduced Young's modulus for joints produced at 800°C. The thin layers close to the base materials revealed a higher hardness value, which is related to the phases constituting them. The central region has a hardness value similar to those of the base materials. The reduced Young's modulus values proved the presence of phases with different elastic properties across the joint, which is in accordance with different modulus of the phases that composed the interface and base materials (Kipp 2010, Toprek et al. 2015).

The joints produced at 800°C attained the highest shear strength value of 225 MPa. This bonding temperature plays a considerable part in eliminating the unbonded regions. However, at the lower temperature of 700°C the shear strength is reduced to 54 MPa, which indicates a weak bond, undetected by microstructural characterisation. These results showed that Ni/Ti reactive multilayers enhance the mechanical strength of dissimilar joints of γ-TiAl

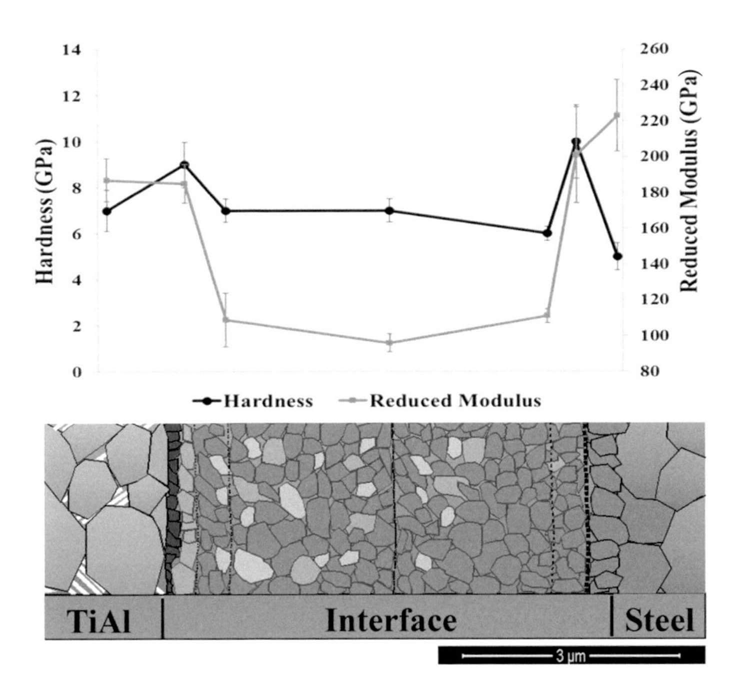

Figure 6.8 Hardness and reduced Young's modulus distributions across a joint produced by diffusion bonding at 800°C with 10 MPa for 60 min.

to steel. The shear strength values are the highest reported for diffusion bonds made in such low processing conditions.

To sum up, the diffusion bonding of γ-TiAl to steel can be improved by using reactive multilayers, such as Ni/Al or Ni/Ti. Reactive multilayers have two beneficial effects: they are effective in reducing bonding conditions, making diffusion bonding an economically more attractive process, and produce joints with higher mechanical properties than when using brazing or diffusion bonding without multilayers. Although the use of Ni/Ti reactive multilayers leads to the formation of a more complex interface, it produces joints with the greatest shear strength values.

6.4 Concluding Remarks

This chapter has described the joining processes and conditions necessary for obtaining a successful bond between γ-TiAl alloys and steel. Brazing, friction welding and diffusion bonding are the processes that have led to the best results. The use of different filler materials has been investigated for brazing γ-TiAl and steel. Sound joints, without brittle intermetallic phases or carbides, are obtained using an Ag-based filler. Nevertheless, the demanding processing conditions and the low creep resistance makes this process unattractive for joining these materials. Friction welding is a valid option for joining γ-TiAl to steel. However, the use of interlayers in the friction process is a requirement for obtaining the desired mechanical properties. Diffusion bonding produces the joints with the highest mechanical properties. Since the required processing conditions hinder its industrial application, a modification of the bonding technique for diffusion bonding of γ-TiAl and steel is required.

The use of reactive multilayers as an interlayer material is a modification of the process that improves diffusion bonding. Different multilayers can be used in bonding γ-TiAl to steel. With Ni/Al reactive multilayers, sound joints were obtained at 800°C for 60 min under a pressure of 50 MPa. The bonding promotes the formation of several interfacial reaction layers with different compositions and thicknesses. These interfacial layers exhibit lower hardness than those formed without the multilayers. The shear strength of the joint produced without multilayers is very low, with an average of 37 MPa, while for the bond produced with the two base materials coated with Ni/Al reactive multilayers, this strength reaches 162 MPa. Ni/Ti reactive multilayers can also be applied to improve diffusion bonding between γ-TiAl and steel. Although these multilayers are less reactive than Ni/Al reactive multilayers, they promote the formation of joints with higher mechanical properties (the average value of the shear strength being 216 MPa).

To sum up, the use of reactive multilayers in diffusion bonding proves to be an effective approach for bonding γ-TiAl to steel. The nanometric multilayers facilitate diffusion across the interface, leading to an effective bond between the multilayers and the base materials and the formation of an interface with less brittle intermetallic phases.

Keywords: Aerospace; automotive; brazing; diffusion bonding; friction welding; fusion welding; interlayers; joining technologies; mechanical properties; microstructure; processing conditions; reactive multilayers; shear strength; steel; tensile strength.

6.5 References

Cao, J., J. Qi, X. Song and J. Feng. 2014. Welding and joining of titanium aluminides. Materials 7: 4930–4962.
Ding, J., J.N. Wang, Z.H. Hu and D.Y. Ding. 2002. Joining of γ-TiAl to low alloy steel by electron beam welding. Mater. Sci. Technol. 18: 908–912.
Dong, H., Z. Yang, G. Yang and C. Dong. 2013. Vacuum brazing of TiAl alloy to 40Cr steel with $Ti_{60}Ni_{22}Cu_{10}Zr_8$ alloy foil as filler metal. Mater. Sci. Eng. A-Struct. Mater. Prop. Microstruct. Process 561: 252–258.
Dong, H.-G., L.-Z. Yu, H.-M. Gao, D.-W. Deng, W.-L. Zhou and C. Dong. 2014. Microstructure and mechanical properties of friction welds between TiAl alloy and 40Cr steel rods. Trans. Nonferrous Met. Soc. China 24: 3126–3133.
Dong, H., Z. Yang, Z. Wang, D. Deng and C. Dong. 2015a. CuTiNiZrV amorphous alloy foils for vacuum brazing of TiAl alloy to 40Cr steel. J. Mater. Sci. Tecnol. 31: 217–222.
Dong, H., L. Yu, D. Deng, W. Zhou and C. Dong. 2015b. Direct friction welding of TiAl alloy to 42CrMo steel rods. Mater. Manuf. Process 30: 1104–1108.
Han, W. and J. Zhang. 2001. Diffusion bonding between TiAl based alloys and steels. J. Mater. Sci. Tecnol. 17: 191–192.
He, P., J.C. Feng, B.G. Zhang and Y.Y. Qian. 2002a. Microstructure and strength of diffusion-bonded joints of TiAl base alloy to steel. Mater. Charact. 48: 401–406.
He, P., J.C Feng, Y.Y. Qian and B.G. Zhang. 2002b. Microstructure and strength of TiAl/40Cr joint diffusion bonded with vanadium-copper filler metal. Trans. Nonferrous Met. Soc. China 12: 811–813.
He, P., J.C. Feng, B.G. Zhang and Y.Y. Qian. 2003. A new technology for diffusion bonding intermetallic TiAl to steel with composite barrier layers. Mater. Charact. 50: 87–92.
He, P., J.C. Feng and W. Xu. 2006. Mechanical property of induction brazing TiAl-based intermetallics to steel 35CrMo using AgCuTi filler metal. Mater. Sci. Eng. A-Struct. Mater. Prop. Microstruct. Process. 418: 45–52.
Kipp, D.O. 2010 Metal Material Data Sheets, MatWeb, LLC. http://www.matweb.com/
Lee, W.-B., Y.-J. Kim and S.-B. Jung. 2004. Effects of copper insert layer on the properties of friction welded joints between TiAl and AISI 4140 structural steel. Intermetallics 12: 671–678.
Li, Y., J.-C. Feng and P. He. 2005. Vacuum brazing of TiAl to 42CrMo steel with Ag-Cu-Ti filler metal. Trans. Nonferrous Met. Soc. China 15: 331–334.
Li, Y., P. He and J. Feng. 2006. Interface structure and mechanical properties of the TiAl/42CrMo steel joint vacuum brazed with Ag-Cu/Ti/Ag-Cu filler metal. Scripta Mater. 55: 171–174.
Liu, H. and J. Feng. 2002. Vacuum brazing TiAl-based alloy to 40Cr steel using Ag-Cu-Zn filler metal. J. Mater. Sci. Lett. 21: 9–10.

Morizono, Y., M. Nishida, A. Chiba, T. Yamamuro, Y. Kanamori and T. Terai. 2004. Diffusion bonding of TiAl alloy to eutectoid steel and its interfacial self-destruction behavior. Mater. Trans. 45: 527–531.
Noda, T., T. Shimizu, M. Okabe and T. Iikubo. 1997. Joining of TiAl and steels by induction brazing. Mater. Sci. Eng. A-Struct. Mater. Prop. Microstruct. Process. 239-240: 613–618.
Simões, S., F. Viana, A.S. Ramos, M.T. Vieira and M.F. Vieira. 2016a. Reaction-assisted diffusion bonding of TiAl alloy to steel. Mater. Chem. Phy. 171: 73–82.
Simões, S., A.S. Ramos, F. Viana, M.T. Vieira and M.F. Vieira. 2016b. Joining of TiAl to steel by diffusion bonding with Ni/Ti reactive multilayers. Metals 6: 96-1-11.
Toprek, D., J. Belosevic-Cavor and V. Koteski. 2015. *Ab initio* studies of the structural, elastic, electronic and thermal properties of $NiTi_2$ intermetallic. J. Phys. Chem. Solids 85: 197–205.
Xu, Q., M.C. Chaturvedi and N.L. Richards. 1999. The role of phase transformation in electron-beam welding of TiAl-based alloys. Metall. Mater. Trans. A-Phys. Metall. Mater. Sci. 30A: 1717–1726.

CHAPTER 7

Joining of γ-TiAl Alloys to Superalloys

7.1 Introduction

Superalloys, particularly, nickel-based superalloys combine good mechanical strength with excellent resistance to creep, oxidation and corrosion. These properties allow their application in components subjected to high service temperatures and aggressive chemical environments, such as components of turbine engines for marine and aerospace industries. The application of these alloys in large structural components also requires an efficient joining process for maintaining high mechanical properties, including at high temperatures. Moreover, successful joining of these alloys to γ-TiAl-based intermetallics allows the production of more complex components with impact in high temperature aerospace applications.

The joining of superalloys to γ-TiAl alloys is difficult by conventional techniques, such as fusion welding processes. Brazing and diffusion bonding are selected as the joining processes capable of producing higher quality joints. The main advantages of these processes consist in avoiding undesirable phase transformations and eliminate segregation, solidification cracking, and distortion stresses, problems commonly encountered in fusion welding. These processes require high temperatures and/or pressures to produce sound bonds between these two materials. The processing conditions can induce microstructural changes impairing the mechanical properties of the base materials. Thus, it is required that the bonding of these materials should be done in less demanding conditions which besides makes the process more appealing from an industrial point of view.

Reaction assisted diffusion bonding using reactive multilayers can also be applied to obtain successful bonds between Ni-based superalloys and γ-TiAl alloys. As in the cases described in previous chapters, this approach allows the production of sound joints with lower bonding conditions, especially the temperature and the pressure, and with the formation of a thin and uniform interface. This chapter will briefly present the processes and processing conditions for joining superalloys to γ-TiAl alloys. The use of reactive multilayers in the diffusion bonding deserves a special attention. The microstructural and mechanical characterizations of the joints produced will be described.

7.2 Joining of γ-TiAl Alloys to Superalloys

The joining of γ-TiAl to superalloys is particularly interesting because of the high mechanical properties of both materials at elevated temperatures. The successful joining of these two materials could lead to the production of more complex components with exclusive mechanical properties that can meet some special requirements for aerospace components. The mass reduction and improved engine efficiency can be achieved by implementing a bonding process to successfully join these two materials (Donachie and Donachie 2002, Reed 2006).

The joining of superalloys is an exciting challenge. The superalloys, especially those possessing higher mechanical properties, are strengthened by a solid solution, by carbide or intermetallic precipitation, and by oxide dispersion. Their joining or repair, particularly when processes involving melting of the base material are used, partially eliminates the effects of the strengthening treatment and produces weak joints, which needs Post-Bond Heat Treatments (PBHT) (Donachie and Donachie 2002, Reed 2006). In addition, for instance, the Ni-based superalloys are susceptible to hot cracking, such as solidification cracking, liquation cracking and ductility-dip cracking, and to strain-age cracking, in the Heat-Affected Zone (HAZ), during PBHT. Thus, fusion welding processes, including laser welding and electron beam welding, need to be selected to take into account the type of superalloy. These joining processes can only be applied to superalloys with a low Ti + Al content, and cannot be used without an appropriate Pre-Bond Heat Treatment (Shinozaki 2001, Khorunov and Maksymova 2013). Brazing and diffusion bonding are the most suitable processes for joining and repairing Ni-based superalloys, since they can avoid the main problems of fusion welding. Several studies regarding the brazing of superalloys were reported (see, for example (Yeh and Chuang 1997, Wu et al. 2000, Khorunov and Maksymova 2013)), while the joining of superalloys by diffusion bonding was less thoroughly investigated (Shirzadi and Wallach 2004, 2005, Guoge et al. 2006, Han et al. 2007).

One of the most important challenges in brazing superalloys is the selection of the filler materials. The role of the brazing filler material is demanding, since it must provide wettability of the surface covered by a resistant oxide, and at the same time ensure the production of joints with high mechanical properties, similar to those of the base material. The best results were achieved using innovative filler materials, such as amorphous alloys or composite filler metals (composed of a suitable filler metal and a filling agent) in which elements such as boron and silicon were added to act as melting point depressants (Yeh and Chuang 1997, Wu et al. 2000, Khorunov and Maksymova 2013).

Little research into the diffusion bonding of Ni-based superalloys has been reported. Gouge et al. (2006) have analysed the microstructure of the interface of Inconel 718 (Ni-19.0Cr-18.5Fe-4.97Nb-0.50Al-3.00Mo-0.90Ti wt.%) joined by diffusion bonding at temperatures ranging from 910 to 1,060°C, for 60 min under a pressure of 32 MPa. The authors reported the formation of the δ-Ni_3Nb phase at lower temperatures (below 1,000°C) and at slower cooling rates (a cooling rate faster than 5°C/min was required to avoid the formation of this phase). The nucleation of this phase must be avoided, since it causes deterioration in the mechanical properties of the joint. Han et al. (2007) proved that the diffusion bonding of this alloy could be improved using a 25 μm thick nickel foil interlayer. Shirzadi and Wallach (2004, 2005) developed a new procedure to remove the oxide layer before diffusion bonding of superalloys and aluminum alloys, which is based on grinding the mating surfaces with abrasive paper containing a small amount of liquid gallium. When this new method was applied to the diffusion bonding of an Inconel 600 alloy (Ni-16.2Cr-9.2Fe-1.1Co-0.6Mn-0.5Si-0.4Cu, wt.%) at a temperature of 1,150°C for 60 min under 8.7 MPa, almost invisible bond line and interfaces with compositions very close to the base materials were produced (Shirzadi and Wallach 2005).

The dissimilar joining of superalloys is an even more challenging task, and a very demanding processing requirements are needed to produce sound joints. As for the similar joints, diffusion bonding and brazing are the most suitable processes for the successful joining of superalloys to other materials, in particular to γ-TiAl alloys, which is the main focus of this chapter.

Sequeiros et al. (2013) investigated the active brazing of the γ-TiAl alloy (Ti-47Al-2Cr-2Nb at.%) to Inconel 718 (Ni-19Cr-19Fe wt.%) using Incusil-ABA (Ag-27.25Cu-12.5In-1.25Ti wt.%) as filler metal. The joining experiments were performed at 730, 830 and 930°C for 10 min. Sound joints, without pores or cracks, were observed for all conditions. The microstructural characterisation revealed a multilayered interface composed of silver and copper solid solutions ((Ag) and (Cu)), $AlNi_2Ti$ and $AlCu_2Ti$. The reaction layers formed near the base materials were essentially composed of intermetallic compounds ($AlNi_2Ti$ and $AlCu_2Ti$ close to the Inconel and γ-TiAl alloys, respectively). The interface is harder than both base materials, with the exception of the central region, consisting of (Ag) and (Cu) solid solutions. The highest shear strength value (228 MPa) was obtained for the joints processed at 730°C. The decrease in shear strength with the corresponding increase in the processing temperature was associated with the increased thickness of the brittle reaction layers formed close to the base materials.

Guedes et al. (2010) investigated the influence of different filler alloys in the brazing of γ-TiAl alloy (Ti-47Al-2Cr-2Nb at.%) to Inconel 718 (Ni-19Cr-19Fe wt.%). The filler alloys used were Incusil-ABA (Ag-27.25Cu-12.5In-1.25Ti wt.%), Nicusil-3 (Ag-28.1Cu-0.75Ni wt.%) and TiNi 67 (Ti-33Ni wt.%). All filler metals allowed the formation of sound joints; in all cases, the filler alloy induced the formation of a multilayered interface. The lowest temperatures at which brazing was successfully performed were 730, 810 and 1,200°C for Incusil-ABA, Nicusil-3 and TiNi 67, respectively. The two Ag-based braze alloys (Incusil-ABA and Nicusil-3) produced an interface with a microstructure and hardness profile similar to described by Sequeiros et al. (2013). Apparently, the use of Incusil-ABA was advantageous, since it enabled the dissimilar joining to be carried out at a lower temperature. Brazing using TiNi alloy as filler produced an interface harder than the base materials; the layer with the highest hardness value (738 ± 18 HV) formed close to the γ-TiAl and was associated with the formation of the AlNiTi intermetallic compound. The use of the TiNi alloy affords advantages over Ag-based alloys, since the absence of low melting point phases allows the joints to be used at higher service temperatures. However, the processing conditions required to braze with the TiNi 67 filler alloy are much more demanding than those for the Ag-based filler alloys, and the high temperature required (1,200°C) can affect the base materials, compromising its application.

In this study, Guedes et al. (2010) also demonstrated that dissimilar joining of γ-TiAl to Ni-based superalloys is more difficult than the similar joining of γ-TiAl. Indeed, sound joints of γ-TiAl were produced using TiNi 67 at temperatures of 1,000°C (Guedes et al. 2004), which is a much lower temperature than that needed for dissimilar joining.

The above studies proved that brazing is a bonding process that permits the joining of γ-TiAl and superalloys successfully. However, it has some disadvantages that make it unsuitable for industrial implementation. Indeed, the interfaces can show some low melting temperature phases that limit the maximum service temperature (due to the composition of the filler metal that enables brazing at lower temperatures), or the processing conditions necessary to obtain a joint with superior mechanical properties can be too demanding and can affect the base materials. In this context, diffusion bonding is a suitable option for the production of joints with a balanced compromise between mechanical properties and processing conditions.

Zhou et al. (2012) investigated the joining of γ-TiAl (Ti-46Al-6(Cr,Nb,Si,B) at.%) to an Fe-based superalloy (GH2036 (Fe-11.5-13.5Cr-7.0-9.0Ni-7.5-9.5Mn-1.10-1.40Mo-1.25-1.55V wt.%)) by diffusion bonding.

The results revealed that even in higher bonding conditions (1,000°C for 60 min under a pressure of 20 MPa), a weak interface (16 MPa of shear strength) with pores was produced. Xiansheng et al. (2015) studied the diffusion bonding of γ-TiAl (Ti-42Al-8Nb-0.2W-0.1B-0.1Y at.%) to an Ni-based superalloy (Ni-25.17Cr-6.46W-0.84Mo-1.10Al-0.54C at.%) in similar processing conditions: 1,000°C for 60 min under a pressure of 50 MPa. The bonded joints exhibited microcracks, and the presence of several phases, $AlNi_2Ti$, Ni_3Ti and γ-TiAl, was detected at the interface. These results demonstrate that diffusion bonding of Ni-based superalloys to γ-TiAl alloys is difficult, even when very demanding processing conditions are applied. The interfaces consisted of brittle intermetallic phases and showed some defects, such as pores and cracks, which impaired the mechanical response of the joints.

To improve the bonding process and obtain interfaces of better quality and with better mechanical properties, some authors have reported the use of an interlayer for joining superalloys to γ-TiAl alloys. Zhou et al. (2012) demonstrated that using Ti-Zr-Cu-Ni alloy as an interlayer enabled the joining of a γ-TiAl to an Ni-based superalloy (GH3536 (Ni-20.5-23.0Cr-17.0-20.0Fe-8.0-10.0Mo-0.5-2.5Co-0.20-1.00W at.%)) at 935°C for 60 min under a pressure of 3 MPa. These dissimilar joints were achieved using less stringent processing conditions than those used to diffusion bond the same γ-TiAl alloy to an Fe-based superalloy, described above, and produced a joint with much higher shear strength (125 MPa).

The combination of titanium and copper foils, with thicknesses of 50 μm and 30 μm respectively, for use as an interlayer, also proved to be a valuable option in joining γ-TiAl to superalloys (Xiansheng et al. 2015). The foils promoted diffusion and reaction, leading to the formation of a more complex interface with different solid solutions and intermetallic compounds, such as α2-Ti3Al, NiTi2, γ-TiAl, rich Cu-AlNiTi and $AlNi_2Ti$. However, the cracks observed in the joints processed without the interlayers were eliminated.

Li et al. (2012) used a titanium interlayer to join an Ni-based superalloy (Ni-20.63Cr-7.46Co-5.44Al-3.03W-2.98Mo-1.96Ti-0.80Si-0.42V at.%) and γ-TiAl (Ti-52.82Al-1.20Cr-1.10V at.%) alloy at temperatures ranging from 960 to 1,040°C and for dwell times up to 30 min. In these joints the interface was also composed of solid solutions and intermetallic compounds. The highest shear strength value measured was 258 MPa for bonds processed at 1,000°C for 10 min. At lower temperatures the reaction between the interlayer and the base materials was insufficient to produce sound joints. The increase in the bonding temperature or dwell time led to an increase in the total thickness of the interface and to the formation of a brittle intermetallic layer which had reduced the shear strength.

7.3 Joining of γ-TiAl Alloys to Superalloys with Reactive Multilayers

As presented in the previous chapters, diffusion bonding of γ-TiAl alloys using reactive multilayers is a very interesting approach for promoting the formation of a stronger interface, while at the same time allowing bonding with lower processing conditions (temperature and pressure). Ni/Al is the reactive multilayer system, which has the strongest exothermic reaction and has been used for the successful diffusion bonding of γ-TiAl to γ-TiAl and γ-TiAl to steel.

The research of Ramos et al. (2009, 2010) showed that the diffusion bonding of an Ni-based superalloy (Ni-20.0Cr-17.9Fe-4.0Nb-1.6Mo-1.2Ti at.%) and γ-TiAl (Ti-50.4Al at.%) was enhanced by the use of Ni/Al multilayers with bilayer thicknesses of 5 and 14 nm. The interface produced with multilayers with the smaller bilayer thickness (5 nm) exhibited some cracks and unbonded regions at the edge of the joint, while the use of multilayers with 14 nm of bilayer thickness led to a more completely-bonded interface. The interdiffusion across the interface promoted the formation of several layers whose composition corresponded to: AlNiTi and $AlNi_2Ti$ phases close to the γ-TiAl side; Ni-rich aluminide and σ-phase close to the Inconel alloy. The hardness of the diffusion layers suggested the formation of hard intermetallic phases.

The microstructure of the diffusion bonded interface between Inconel 718 (Ni-19.49Cr-18.27Fe-3.51Nb-2.25Al-2.03Mo-1.50Ti at.%) and γ-TiAl (Ti-45.0-5.0Nb at.%), at 800°C, for 60 min and under a pressure of 5 MPa, using Ni/Al multilayers can be observed in Figure 7.1. The interface region is

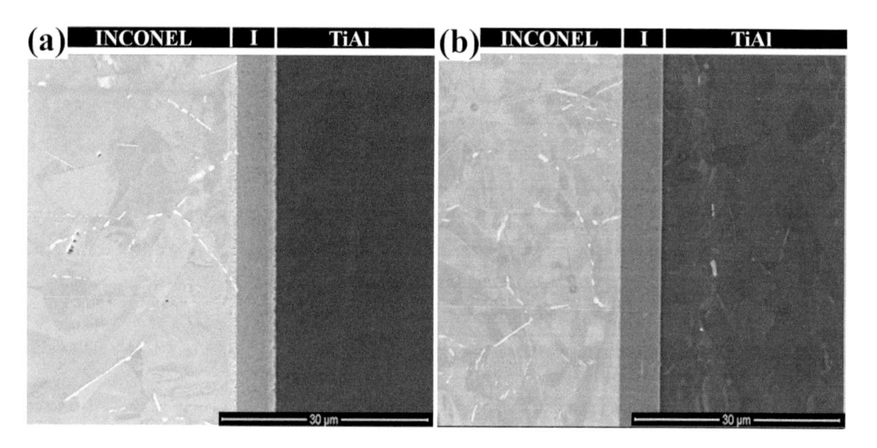

Figure 7.1 SEM images of diffusion bond interfaces between γ-TiAl and Inconel obtained at 800°C under a pressure of 5 MPa for 60 min, using Ni/Al reactive multilayers with bilayer thicknesses of: (a) 5 nm and (b) 14 nm.

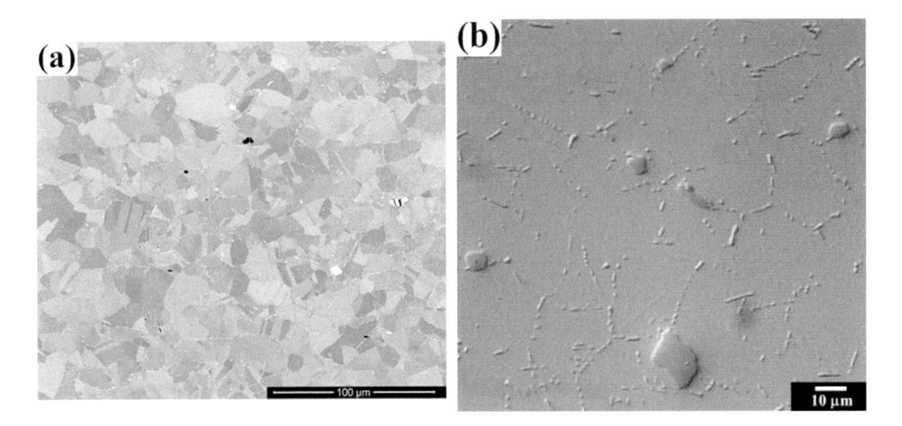

Figure 7.2 (a) SEM and (b) Optical Microscopy (OM) images of the Inconel alloy.

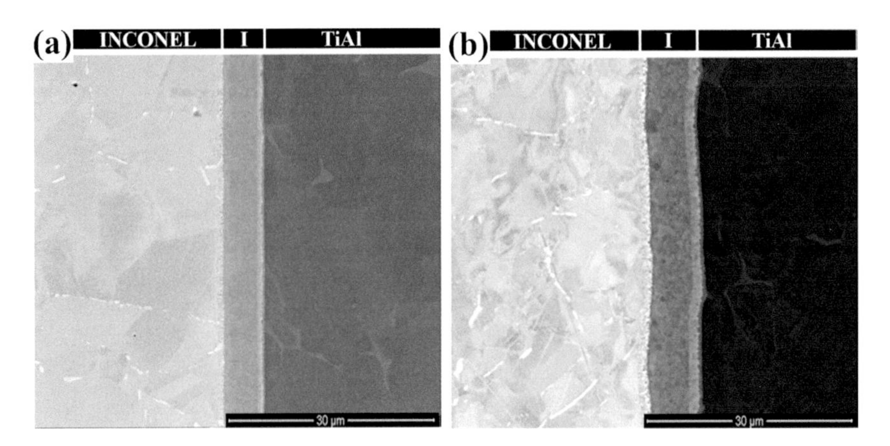

Figure 7.3 SEM images of the diffusion bond interfaces between Inconel 718 and γ-TiAl alloys obtained using multilayers with 14 nm of bilayer thickness, processed for 60 min under a pressure of 5 MPa at: (a) 700°C and (b) 900°C.

clearly visible in these Scanning Electron Microscopy (SEM) images due to the different microstructures of the base materials. It can be observed that a joint with a thin interface (5 µm) is produced and that the microstructure of the base materials remained unchanged (for comparison, the Inconel microstructure is shown in Figure 7.2). Other joining experiments were conducted at 700 and 900°C, using the same dwell time and pressure. SEM images of Figure 7.3 revealed the interface of these joints. Plastic deformation of Inconel occurred during bonding at 900°C, visible in the image in Figure 7.3 (b), indicating that this temperature was too high for bonding. Diffusion bonding at 700°C was possible, but cracks and unbonded regions were observed at the edges of the joint.

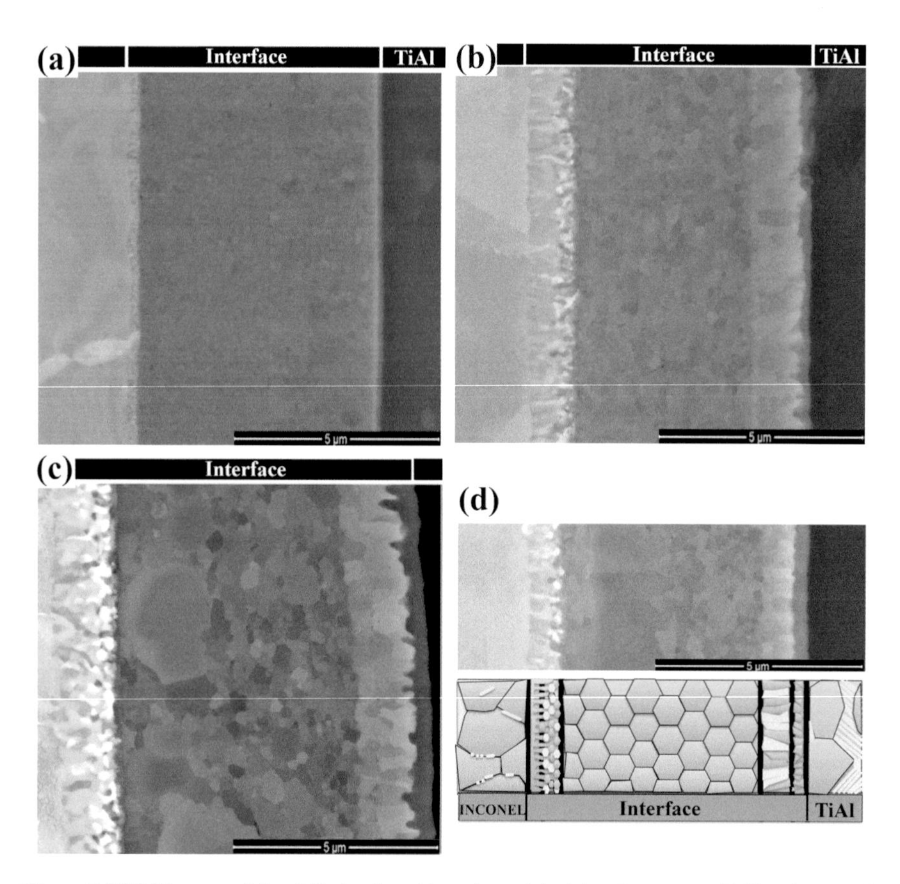

Figure 7.4 SEM images of the diffusion bond interface of the joints between γ-TiAl and Inconel 718 obtained using multilayers with a bilayer thickness of 14 nm processed at: (a) 700°C, (b) 800°C and (c) 900°C and (d) schematic illustration of the layers that constitute the interface processed at 800°C.

Figure 7.4 presents magnified images of the interfaces produced by bonding at 700, 800 and 900°C, and a schematic illustration of the regions detected at the interfaces. The morphology of the interfaces was very similar for all processing conditions: two thin layers formed close to both base materials and a central region composed of equiaxed grains. The main difference between the interfaces was their thickness, which increased with the bonding temperature. Diffusion between the Ni/Al multilayers and the base materials seemed to play a key role in the diffusion bonding. The central region, composed of equiaxed grains, was comparable to the ones produced with the same multilayer system in similar γ-TiAl bonds as well as in dissimilar γ-TiAl to stainless steel bonds (as mentioned in Chapters 5 and 6). The two thin layers close to the γ-TiAl base material were also detected in the dissimilar γ-TiAl to stainless steel bonds (Chapter 6). The formation of a thin layer enriched in nickel close to the γ-TiAl is a unique feature of the two dissimilar bonds (to stainless steel and to Inconel). This layer forms only if the nickel content of the interface is high, as is the case with the interfaces formed from stainless steel and Inconel. The two thin layers close to the Inconel resulted from the diffusion between Inconel and Ni/Al reactive multilayers.

The Energy-Dispersive X-Ray Spectroscopy (EDS) analysis of the bonding interface processed at 800°C (Figure 7.5) confirms the occurrence of an intense diffusion. Indeed, iron, chromium, nickel and aluminum (elements from the superalloy and multilayers) are detected close to Inconel, while titanium, nickel and aluminum (from γ-TiAl and multilayers) are detected close to the γ-TiAl alloy. EDS and Electron Backscatter Diffraction (EBSD) results revealed that the chemical composition and crystallographic structure of the equiaxed grains at the central region of the interface match those of the NiAl phase, which results from the reaction of the Ni/Al multilayer. Close to the γ-TiAl, the bright layer was identified as $AlNi_2Ti$ and the darker one comprised alternating grains of AlNiTi and Al_2NiTi. Moreover, this layer was formed by interdiffusion between the γ-TiAl alloy and the Ni/Al multilayers and is always detected at these interfaces. The Ni_3Al, NiAl and σ-phase (an iron and chromium rich phase) grains comprised the layers closest to the Inconel. Figure 7.6 shows two Kikuchi patterns from these regions that were indexed as NiAl and Ni_3Al phases. The formation of phases such as the δ-phase, reported by some authors as detrimental to the mechanical properties of the joints, was not detected.

The hardness of the interface was evaluated by nanoindentation tests. As in similar joints, it was necessary to conduct several tests in order

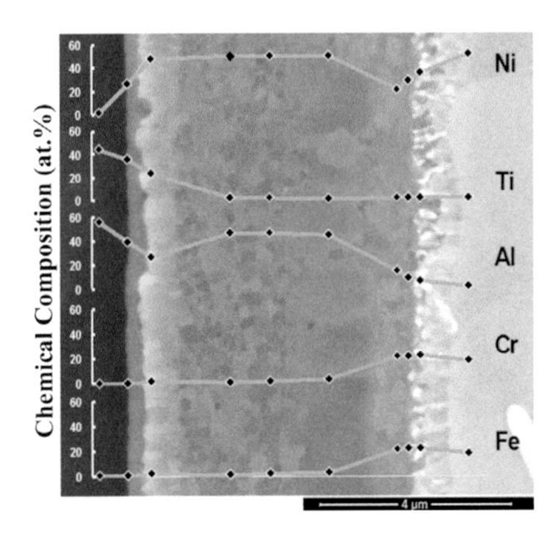

Figure 7.5 SEM image of the bond interface of joints between γ-TiAl and Inconel obtained using multilayers with 14 nm of bilayer thickness at 800°C and the variation of the elemental composition across the interface determined by EDS analysis.

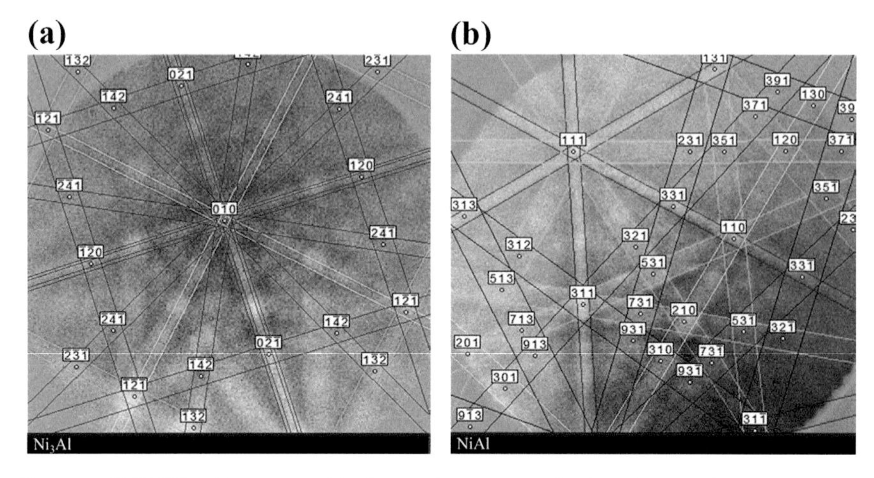

Figure 7.6 EBSD Kikuchi patterns of the layers close to Inconel indexed as Ni_3Al (a) and NiAl (b).

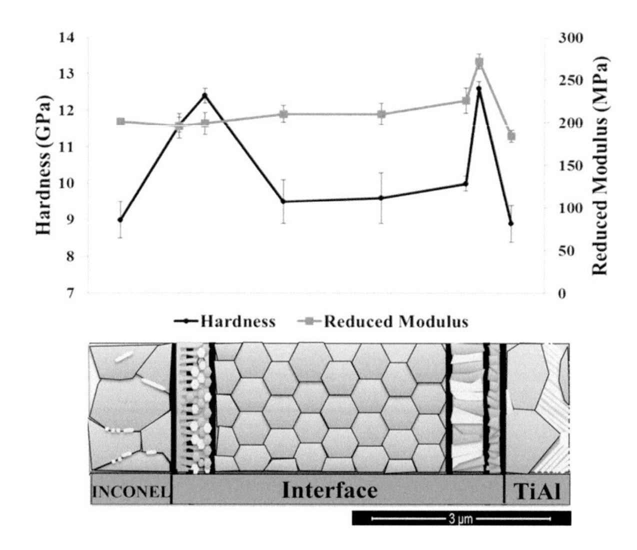

Figure 7.7 Hardness and reduced Young's modulus evolution across the diffusion bond processed at 800°C.

to obtain a number of indentations sufficient to determine the average hardness of each layer, especially for the thinner layers closest to the base materials. Figure 7.7 shows the evolution of the hardness across the interface. The thin layers which formed close to both base materials, composed of intermetallic brittle phases, revealed the highest hardness values, while the central zone exhibited a hardness value close to those of the base materials.

To sum up, the use of Ni/Al reactive multilayers with 14 nm of bilayer thickness enhanced the diffusion bonding of the Ni-based superalloy and γ-TiAl. The use of these multilayers enabled the production of sound joints at a low temperature (800°C) and under a lower pressure (5 MPa) than those used in conventional diffusion bonding with and without interlayers (900–1,000°C and 20 MPa). Especially for Inconel, it is very important to establish joining processes at lower temperatures and pressures in order to avoid the unwanted structural changes that compromise the mechanical properties of the components.

7.4 Concluding Remarks

This chapter has described the joining processes and conditions for bonding γ-TiAl alloys and superalloys. Fusion welding processes cannot be applied in all superalloys; for instance, Ni-based superalloys are susceptible to hot cracking and to strain-age cracking during PBHT. Brazing and diffusion bonding are shown to be the best joining processes for the successful production of these joints. Brazing using Ag-based filler allows the production of sound dissimilar joints, but the formation of low melting temperature phases at the interface compromises the use of these joints at high temperatures. Other fillers result in a significant increase in the brazing temperature. Diffusion bonding is a process that produces sound dissimilar joints with the desired properties using high temperatures and/or pressures. The bonding of these materials in less demanding conditions and with a thin interface is required.

Diffusion bonding assisted by reactive multilayers is one option for the production of joints between Ni-based superalloys and γ-TiAl alloys. This approach permits the production of sound joints under less demanding bonding conditions, mainly at lower temperature and pressure, and the formation of a thin and more uniform interface. Ni/Al reactive multilayers with 14 nm of bilayer thickness enhances diffusion bonding, producing sound joints at 800°C for 60 min under a pressure of 5 MPa. The bond between the two base materials is ensured by the diffusion between the nanometric multilayers and the elements of the base materials. The interface consists of a central region composed of NiAl equiaxed grains, which corresponds to 90 per cent of the interface, and of thin layers close to the base materials composed of Al-Ni-Ti intermetallic compounds, Ni_3Al, NiAl and σ grains.

Keywords: Aerospace; brazing; diffusion bonding; fusion welding; hardness; interface; interlayers; joining technology; mechanical properties; microstructure; reactive multilayers; superalloys.

7.5 References

Donachie, M.J. and S.J. Donachie. 2002. Superalloys: A Technical Guide, 2nd Edition. ASM International, Materials Park, OH, USA.

Guedes, A., A.M.P. Pinto, M.F. Vieira and F. Viana. 2004. Joining Ti-47Al-2Cr-2Nb with a Ti-Ni braze alloy. Mater. Sci. Forum. 455-456: 880–884.

Guedes, A., H. Mollaoglu, F. Viana, A.M.P. Pinto and M.F. Vieira. 2010. Brazing Ti-47Al-2Cr-2Nb to Inconel 718 with different filler alloys: Microstructural characterization of the interfaces. Proc. of the Int. Conf. on Advances in Welding Science and Technology for Construction, Energy and Transportation, AWST 2010, held in Conj. with the 63rd Annual Assembly of IIW 2010. Istanbul, Turkey. 353–359.

Guoge, Z., R.S. Chandel, S.H. Pheow and H.H. Hoon. 2006. Effect of bonding temperature on the precipitation of δ phase in diffusion bonded Inconel 718 joints. Mater. Manuf. Process. 21: 453–457.

Han, W.B., K.F. Zhang, B. Wang and D.Z. Wu. 2007. Superplasticity and diffusion bonding of IN718 superalloy. Acta Metall. Sin. 20: 307–312.

Khorunov, V.F. and S.V. Maksymova. 2013. Brazing of superalloys and the intermetallic alloy (γ-TiAl). pp. 85–120. *In*: D.P. Sekulic [ed.]. Advances in Brazing Science, Technology and Applications. Woodhead Publishing India Private Limited, New Delhi, India.

Li, H.-X., P. He, T.-S. Lin, F. Pan, J.-C. Feng and Y.-D. Huang. 2012. Microstructure and shear strength of reactive brazing joints of TiAl/Ni-based alloy. Trans. Nonferrous Met. Soc. China. 22: 324–329.

Ramos, A.S., M.T. Vieira, S. Simões, F. Viana and M.F. Vieira. 2009. Joining of superalloys to intermetallics using nanolayers. Adv. Mater. Res. 59: 225–229.

Ramos, A.S., M.T. Vieira, S. Simões, F. Viana and M.F. Vieira. 2010. Reaction-assisted diffusion bonding of advanced materials. Defect. Diffus. Forum. 297-301: 972–977.

Reed, R.C. 2006. The Superalloys: Fundamentals and Applications. Cambridge University Press, New York, USA.

Sequeiros, E.W., A. Guedes, A.M.P. Pinto, M.F. Vieira and F. Viana. 2013. Microstructure and strength of γ-TiAl alloy/Inconel 718 Brazed Joints. Mater. Sci. Forum. 730-732: 835–840.

Shinozaki, K. 2001.Welding and joining Fe and Ni-base superalloys. Weld. Inter. 15: 593–610.

Shirzadi, A.A. and E.R. Wallach. 2004. New method to diffusion bond superalloys. Sci. Technol. Weld Joi. 9: 37–40.

Shirzadi, A.A. and E.R. Wallach. 2005. Novel method for diffusion bonding superalloys and aluminium alloys (USA Patent 6,669,534 B2, European Patent Pending). Mater. Sci. Forum. 502: 431–436.

Wu, X., H. Li, R.S. Chandel, F. Lan and H.P. Seow. 2000. Effect of mechanical vibration on TLP brazing with BNi-2 nickel-based filler metal. J. Mater. Sci. Lett. 19: 319–321.

Xiansheng, Q., X. Xiangyi, T. Bin, K. Hongchao, H. Rui and L. Jinshan. 2015. Phase evolution of diffusion bonding interface between high Nb containing TiAl alloy and Ni-Cr-W superalloy. Rare Metal Mater. Eng. 44: 1575–1580.

Yeh, M.S. and T.H. Chuang. 1997. Effects of applied pressure on the brazing of superplastic Inconel 718 superalloy. Metall. Mater. Trans. A 28: 1367–1376.

Zhou, Y., H.-P. Xiong, W. Mao, B. Chen and L. Ye. 2012. Microstructures and property of diffusion bonded joints between TiAl alloy and two kinds of superalloys. J. Mater. Eng. 8: 88–91.

CHAPTER 8

Combination of Reactive Multilayers and Brazing Alloys for Improving Joining Processes

8.1 Introduction

Throughout this book, several studies that demonstrate the effective use of reactive multilayers as interlayers to improve diffusion bonding of γ-TiAl alloys have been presented. In similar and dissimilar joints, reactive multilayers, particularly those of Ti/Al, Ni/Al and Ni/Ti, can effectively decrease the temperature and pressure of diffusion bonding processes, producing very thin interfaces with a mechanical strength similar to that of the base materials. In dissimilar joints, these multilayers can also act as a diffusion barrier to prevent the formation of harmful phases.

Brazing of γ-TiAl alloys with Ti-Cu-Ni or Ti-Ni brazing alloys has also been employed to produce sound joints, thus preventing some problems that occur in the fusion welding processes, such as thermal shock and high residual stresses. Brazing is also able to overcome shortcomings in diffusion bonding processes, and fill the bond gap, as well as cracks and other imperfections in the mating surfaces. However, the formation of a multiphase interface, with intermetallic compounds different from the base materials, and the consequent variation in mechanical behaviour, cannot be avoided. Additionally, the brazing temperatures are still very high. The decrease in the brazing temperature can be achieved by using silver-based brazing alloys. However, the maximum service temperature of these joints is noticeable inferior, due to the formation of compounds with low melting temperatures.

Modifying the bonding technique, by combining reactive multilayers with brazing alloys, is an interesting solution that has been explored in several studies published in the literature. This procedure may increase the diffusivity of the interface and reduce the bonding temperature.

The combination can be implemented in various ways, by overlapping the multilayers with brazing alloy (the brazing alloy closest to the base material and the multilayers in the center, or the opposite), coating the mating surfaces of the base materials with reactive multilayers and placing the brazing alloy at the center, or coating the multilayers with the brazing alloys and placing the assembly between the mating surfaces. The large amount of heat released by the reaction of the multilayers is used to melt the brazing alloy, acting as a controllable local heat source. After ignition, the reaction is propagated through the multilayer so that the heat is localised and progressively melts the brazing layers. Joining can be performed in any atmosphere or under vacuum, and the temperature of the joined components is never raised significantly.

Ni/Al reactive multilayers are one of the most promising reactive systems to use in combination with brazing alloys, since they exhibit the highest heat release, with higher propagation velocities and the formation of the final phase in a short period of time.

In fact, Ni/Al reactive multilayers coated with brazing alloys are already commercialised as NanoFoil® by Indium Corporation, to be used in joining processes, especially to permit joining at room temperature. However, the key to the success of this combination also depends on the heat release and phase formation during the multilayer reaction, as well as on the diffusion and reaction between the elements of the base materials, multilayers and brazing alloys. The main challenge is to prevent the formation of brittle phases that will compromise the mechanical properties of the joints.

The aim of this chapter is to describe the results of bonding γ-TiAl alloys using the combination of reactive multilayers and brazing alloys. More importance will be given to improvements in the bonding process achieved by this approach. The microstructural features and mechanical properties of the joints will be explored.

8.2 NanoFoil®

The intention in combining reactive multilayers and brazing alloys is to allow bonding at lower temperatures. The multilayers and brazing alloys can be used as freestanding films, but if they are used as coatings, they can eliminate the problems of adhesion between two surfaces of the joints: between base material/multilayers or base material/brazing alloy or multilayers/brazing alloy. NanoFoil® is an Ni/Al reactive multilayer coated with brazing alloys. In some joints, the use of the NanoFoil® also permits bonding at room temperature. However, the success of its application in joining processes is strongly dependent on the base materials.

NanoFoil® is a multilayer of nickel (alloyed with small amounts of vanadium) and aluminum alternated nanolayers produced by sputtering (Indium Corporation 1996–2016). The foils can be used, as produced or combined, with brazing alloys. The foils are available with thicknesses of 40, 60 or 80 µm, and the bilayer thickness is selected according to the heat release requirements. These reactive multilayers provide localised heat of up to 1,500°C in a nanosecond (Indium Corporation 1996–2016). The brazing alloys need to be selected to meet the requirements: to minimize thermal exposure of the base materials and ensure compatibility with the reactive multilayers and with the base materials, while maximizing the thermal conductivity at lower costs. Figure 8.1 displays the Scanning Electron Microscopy (SEM) images and a schematic illustration of a NanoFoil®, coated with an Ag-Cu-In braze alloy with a thickness of 280 nm. This NanoFoil® is characterised by uniform alternating layers of nickel and aluminum with a 54 nm bilayer thickness and a total thickness of 60 µm (Simões et al. 2014). The columnar microstructure is typical of reactive multilayers produced by sputtering.

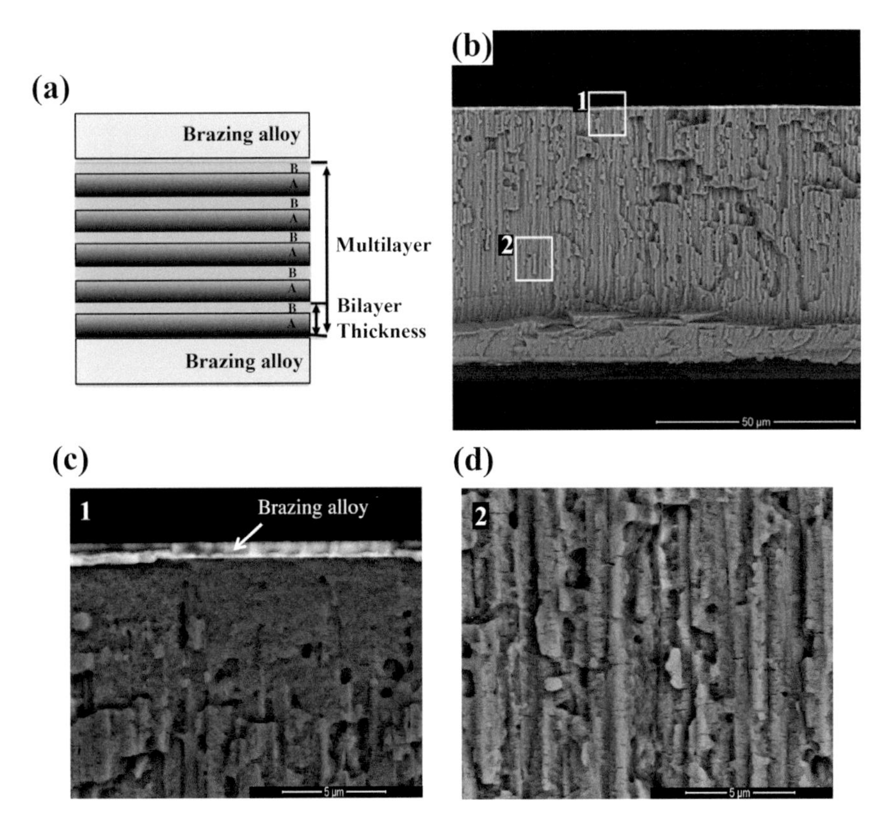

Figure 8.1 NanoFoil® coated with a brazing alloy (a) schematic illustration, (b) SEM image and (c) and (d) higher magnification of the regions marked in (b).

The reaction between the aluminum and nickel nanolayers of the NanoFoil® can be ignited by an electrical discharge (self-propagating reaction) or by slow heating, with a structural evolution strongly dependent on the ignition method. Table 8.1 summarises the results of the phase transformations of NanoFoil®. When the reaction was ignited by an electric discharge, the alternating layers reacted and formed NiAl with a micrometric grain size, as can be seen in the image in Figure 8.2 (a). This reaction was associated with a contraction of approximatively 20 per cent in volume. When the multilayers were annealed, the reaction path depended on the heating rate. For rapid heating rates, 40°C/min, the nanolayers transformed at 204°C directly into NiAl

Table 8.1 Temperature of reactions and phase transformations of Ni/Al multilayers with different chemical composition and bilayer thickness.

Chemical composition	Bilayer thickness (nm)	Ignition	Peak temperature (°C)	Phases	Characterization techniques
Ni 49Al (at.%) with Ag-Cu-In brazing alloy (Simões et al. 2016)	48	Electric discharge	—	NiAl	DSC and XRD
		Annealing 40°C/min	204	NiAl	
		Annealing 10°C/min	180	$NiAl_3$	
			230	Ni_2Al_3	
			320	Ni_3Al_4	
			380	NiAl	
Ni-60Al-3V (at.%) (Trenkle et al. 2010)	100	Annealing	1377	Al_3V + Ni_2Al_3 + NiAl	DSC and *In situ* X-Ray microdiffraction
Ni-49Al-4V (at.%) (Trenkle et al. 2010)	100	Annealing	1567	NiAl	
Al/Ni0.91V0.09 (Al to Ni0.9V0.09 2:3 atomic ratio) (Kim et al. 2008)	10	Annealing	1427	NiAl	Dynamic single-shot diffraction and dynamic TEM
Al to Ni-V 1:1 atomic ratio with Incusil-ABA braze alloy (Ag-27.3Cu-12.5In-1.25Ti wt.%) (Duckham et al. 2004a)	25	Annealing 40°C/min	—	NiAl	DSC and XRD
	80		—	NiAl	

DSC: Differential Scanning Calorimetry; XRD: X-Ray Diffraction; TEM: Transmission Electron Microscopy.

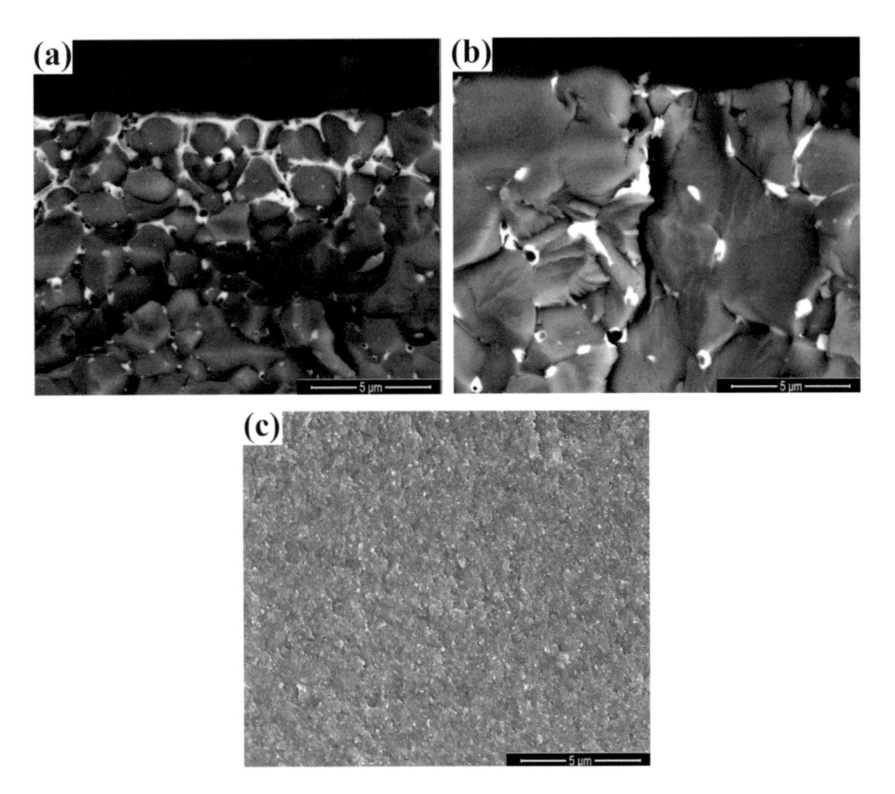

Figure 8.2 SEM images of the fracture surface of a NanoFoil® coated with a brazing alloy: (a) after the reaction triggered by an electrical discharge, (b) after annealing at 500°C with a 40°C/min heating rate and (c) after annealing at 700°C with a 10°C/min heating rate.

with a micrometer grain size, as in the previous case. For slow heating rates, 10°C/min, the reaction path was more complex and occurred in four stages, which are indicated in Table 8.1. Annealing at 700°C with a slow heating rate also promotes the formation of NiAl as the final phase, though with a nanometric grain size of 73 ± 22 nm (Simões et al. 2016). The difference in the grain size of the reaction products is evident when one observes the microstructures in Figure 8.2. This difference is probably the consequence of varied temperatures attained by the multilayers during the reaction; the ignition by an electric discharge promotes a fast reaction and a high heat release rate that, according to the Nanofoil® producers (Indium Corporation 1996–2016), can heat the NanoFoil® by up to 1,500°C in a nanosecond. *In-situ* experiments (Kim et al. 2008, Trenkle et al. 2010) of self-propagating Ni/Al multilayer reactions showed that aluminum melts, thus assisting the intermixing and NiAl nucleation and therefore promoting the formation of larger NiAl grains. For slow heating rates, liquid aluminum was not detected and the reaction occurred in several steps

(Kim et al. 2008). Duckham et al. (2004a) also investigated the exothermic reactions in the NanoFoil® by Differential Scanning Calorimetry (DSC). When the NanoFoil® reaction occurs by annealing with a heating rate of 40°C/min, only one exothermic peak is observed in DSC curves. The temperature of the reaction depends on the bilayer thickness; smaller bilayer thicknesses exhibit lower reaction temperatures. The effect of the annealing temperature on grain size was also demonstrated for other multilayers; heating an Ni/Al multilayer with a bilayer thickness of 30 nm from 450 to 700°C resulted in a grain growth from 22 ± 8 to 116 ± 44 nm (Simões et al. 2011).

In the SEM images, it is possible to observe pores, probably shrinkage porosity formed during the solidification of aluminum at the early stages of the reaction, and/or Kirkendall pores that result from the difference between the aluminum and nickel diffusion fluxes. In fact, the DSC curves exhibit an endothermic peak that corresponds to the melting of the aluminum. Part of the aluminum was spent in the formation of intermetallic phases and the remaining melted, probably due to the eutectic reaction ($Al + NiAl_3 \rightarrow L$) (Simões et al. 2016).

8.3 Joining Processes Assisted by the Combination of Reactive Multilayers and Brazing Alloys

The use of reactive multilayers proved to be an effective approach for enhancing the bonding of electronic components, superalloys and γ-TiAl alloys. The assembly of reactive multilayers and brazing alloys aims to improve the brazing processes, by using the heat of the multilayer reaction to melt the brazing alloy. By means of this approach, the heat source is localised inside the joint and the base materials are not exposed or affected by high temperatures. Additionally, in dissimilar joints, the multilayers can act as a diffusion barrier to limit the formation of undesirable phases. Different configurations can be used for the application of this approach: overlapping the multilayers with brazing alloy, using base materials coated with multilayers and a freestanding brazing alloy foil, or coating the multilayers with the brazing alloys.

Some studies (Wang et al. 2003, 2004, Duckham et al. 2004a,b, Qiu and Wang 2008, Bartout and Wilden 2012) showed that the combination of multilayers and brazing alloys led to the reduction in the bonding temperature and can even allow joining at room temperature. Wang et al. (2004) reported the joining of stainless steel at room temperature using Ni/Al reactive multilayers and an AuSn (Au-20Sn, wt.%) brazing alloy. The Ni/Al reactive multilayers (50/50 atomic ratio of Al and Ni) were deposited by sputtering with bilayer thicknesses ranging from 35 to 85 nm and total foil thicknesses ranging from 70 to 170 μm. Stacks of two 25 μm thick sheets of AuSn solder and one reactive foil between two stainless steel samples were joined under a pressure of approximately 100 MPa. The multilayer reaction was ignited and the heat released melted the brazing alloy, assuring the joining. These joints presented a shear strength of 48 MPa, slightly higher than the strength of conventionally brazed joints, which is 38 MPa. The feasibility of using the Ni/Al multilayers coated with an Incusil ABA (59 Ag–27.25Cu–12.5In–1.25Ti, wt.%) brazing alloy, 1 μm thick, combined with AuSn (Au-20Sn, wt.%) and AgSn (Sn-3.5Ag, wt.%) brazing alloys to bond a 6061 Al alloy and 316L stainless steel, was also investigated (Wang et al. 2004b). The results demonstrated that the strength of joints increased in line with the bonding pressure until a critical value was attained. This critical pressure depended on the base material, the foil thickness and the brazing alloy. If an AuSn brazing alloy and reactive multilayers with a thickness of 40 μm were used, the pressure needed for a sound stainless steel bond was at least 10 kPa. For the Al alloy a higher pressure was required; with a 100 μm multilayer, a pressure of at least 10 MPa was necessary to form a strong joint, with an average maximum shear strength of 35 MPa.

Duckham et al. (2004a) reported that this approach could also be used to join titanium-based alloys. Ti6Al4V was successfully bonded at room temperature and in air using freestanding Ni/Al reactive multilayers and a silver-based brazing alloy. The freestanding Ni/Al reactive multilayers were deposited by sputtering; the Incusil-ABA brazing alloy (Ag-27.3Cu-12.5In-1.25Ti, wt.%) was deposited on the surfaces of the multilayer foil and into the base material. The temperature reached by the multilayers was found to be critical for the success of the joining process when higher melting temperature brazing alloys were used. Other materials were joined using the reactive multilayers and brazing alloys. Qiu and Wang (2008) reported the use of reactive multilayer foils and brazing alloys to bond silicon wafers. Two AuSn (Au-20Sn, wt.%) brazing alloy foils and one freestanding Ni/Al multilayer foil were stacked between the base materials. An electric spark was used to ignite the multilayer reaction. A sound bond was then produced with strength higher than the base material.

This approach can also be applied to bonding materials which are dissimilar and those which are difficult to bond, such as metal/ceramic materials, which are very interesting yet difficult to bond. Soldering and active metal brazing are the most effective processes in bonding these materials. The reactive multilayers are a promising interlayer for achieving a sound bond at low temperature. For instance, Duckham et al. (2004b) reported the joining of Ti-6Al-4V to SiC using reactive multilayers and brazing alloys. Freestanding Ni/Al reactive multilayer foils, Ti-6Al-4V and SiC, were coated with the Incusil-ABA brazing alloy. The bonding experiments were performed at room temperature under a pressure of 35 MPa. Sound joints were obtained, though the success of the bonding depended on the maximum temperature reached during the reaction, reaction velocity, foil thickness and heat released during the reaction. Joining of aluminum to copper has also been performed using Ni/Al multilayers and brazing alloys; Bartout and Wilden (2012) reported sound dissimilar bonds using Ni/Al multilayers and an tin-based brazing alloy. The use of the combination of reactive multilayers and brazing alloys shows enormous potential for joining, ensuring the bond between different materials, reducing the bonding temperature and even enabling the joining to be processed without the requirement of vacuum or inert atmosphere.

Joining lightweight materials such as aluminum and magnesium can also be improved by the use of NanoFoil® (Sun 2010). The NanoFoil® bond strength is comparable to that of the conventional structural adhesive bonds, attesting to the success in the application of this approach to joining components for the automotive industry.

8.4 Joining of γ-TiAl Alloys using the Combination of Reactive Multilayers and Brazing Alloys

The research presented in the above section has shown that bonding processes can be improved by the combination of reactive multilayers and brazing alloys. This approach permits a decrease in the bonding temperature while enabling the melted brazing alloys to fill the gaps between the mating surfaces. Commercial Ni/Al reactive multilayers coated with brazing alloys demonstrate a huge potential for permitting joining at room temperature. In this section, the use of the combination of reactive multilayers with brazing alloys in the joining of γ-TiAl alloys will be described in order to present the advantages and disadvantages of different combinations and configurations.

8.4.1 Joining of γ-TiAl Alloys with NanoFoil®

NanoFoil® can be used to bond γ-TiAl alloys at room temperature under pressures of 20 and 40 MPa. Figure 8.3 shows the microstructure of the interface of the joints processed at room temperature with a 60 μm thick NanoFoil®, coated with an Ag-Cu-In braze alloy 280 nm thick. The multilayer reaction was ignited by an electric discharge under pressure. In the SEM images, a bright zone is visible along the interface between the NanoFoil® and the γ-TiAl alloy (Ti-45Al-5Nb at.%), corresponding to a reaction layer between the coating of the multilayers and the base material. Another feature of these joints was the formation of cracks on the NanoFoil®, perpendicular to the interface. Some of these cracks were filled with the coating material. The number of cracks increased with the bonding pressure. There are two probable causes for the formation of these cracks: the contraction of the multilayers during the reaction, which reaches

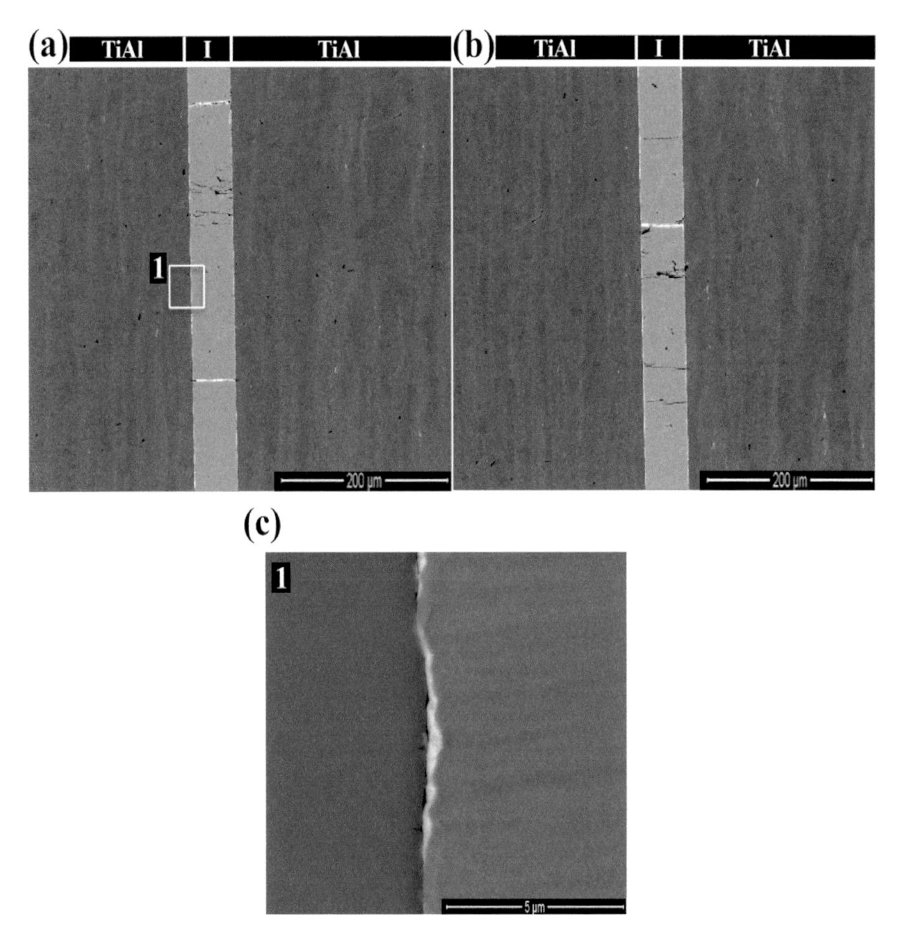

Figure 8.3 SEM images of γ-TiAl alloy joints produced with NanoFoil® at room temperature: (a) under a pressure of 20 MPa and (b) under a pressure of 40 MPa, (c) is a high magnification image of the region 1 indicated in (a).

20 per cent in volume (Simões et al. 2016), or the fracture of the NanoFoil® by the pressure applied during bonding, as a result of the residual stresses and brittleness of the nanolayers prior to the ignition of the reaction.

The interfaces formed can be divided into three distinct zones: two thin layers close to the γ-TiAl alloy and a larger central zone (Figure 8.3). According to the Energy-Dispersive X-Ray Spectroscopy (EDS) analysis, the reaction between γ-TiAl and the Ag-Cu-In coating resulted in the formation of two thin zones, from the γ-TiAl alloy to the center of interface: an Al-Cu-Ti layer and a silver-rich layer. At the center of the interface only NiAl grains were observed, corresponding to the reacted multilayer. Interdiffusion or reaction between the Ag-Cu-In

Table 8.2 Microhardness values of γ-TiAl alloys similar and dissimilar joints produced with NanoFoil®.

Joints	Processing conditions	Zone	Hardness HV 0.01
Similar γ-TiAl	Room temperature 20 MPa	γ-TiAl	417
		Thin layers close to γ-TiAl	540
		Center of interface	926
	Room temperature 40 MPa	γ-TiAl	419
		Thin layers close to γ-TiAl	569
		Center of interface	960
	700°C/60 min/10 MPa	γ-TiAl	414
		Thin layers close to γ-TiAl	450
		Center of interface	1,129
	800°C/60 min/5 MPa	γ-TiAl	415
		Thin layers close to γ-TiAl	475
		Center of interface	1,090
Dissimilar γ-TiAl to Inconel	700°C/60 min/5 MPa	Inconel	419
		Thin layers close to Inconel	561
		Center of interface	1,312

coating and the Ni/Al multilayer was not detected, possibly due to the reduced thickness of the interface. Table 8.2 displays the hardness values of these interfaces. The highest hardness value was determined at the center of the interface, the region with NiAl grains.

Dissimilar joints of γ-TiAl alloys were also performed at room temperature. Figure 8.4 shows the interface obtained between the γ-TiAl alloy (Ti-45Al-5Nb at.%) and an nickel-based superalloy (Inconel718 (Ni-19.0Cr-18.5Fe-4.97Nb-0.50Al-3.00Mo-0.90Ti wt.%)). The main characteristic of these joints is the formation of cracks perpendicular to the interface, as observed for similar γ-TiAl joints. For dissimilar joints, it was necessary to increase the pressure to 80 MPa to eliminate the unbonded areas.

Bonding with NanoFoil® using annealing for the ignition of the multilayer reaction was implemented to prevent the formation of cracks at the joints and to improve the quality of similar and dissimilar γ-TiAl joints. Figure 8.5 shows the interfaces of similar and dissimilar γ-TiAl joints produced at 700°C for 60 min under a pressure of 10 MPa. The interfaces were free of defects and the cracks were not detected. Different zones were observed at the interface (Figure 8.5). A very thin Al-Cu-Ti layer formed close to the γ-TiAl alloy, followed by a thin layer rich in silver. Close to the multilayers a zone rich in nickel and aluminum with some copper and

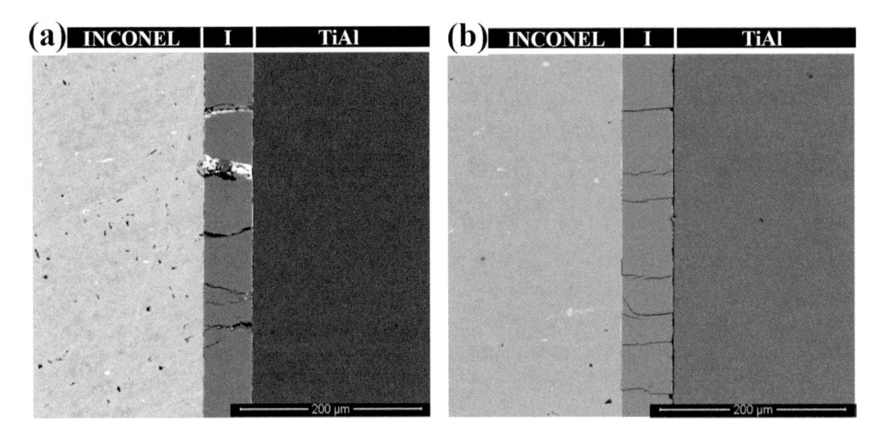

Figure 8.4 SEM images of the interface of the dissimilar joints of γ-TiAl alloy and Inconel produced with NanoFoil® at room temperature, under pressures of (a) 40 MPa and (b) 80 MPa.

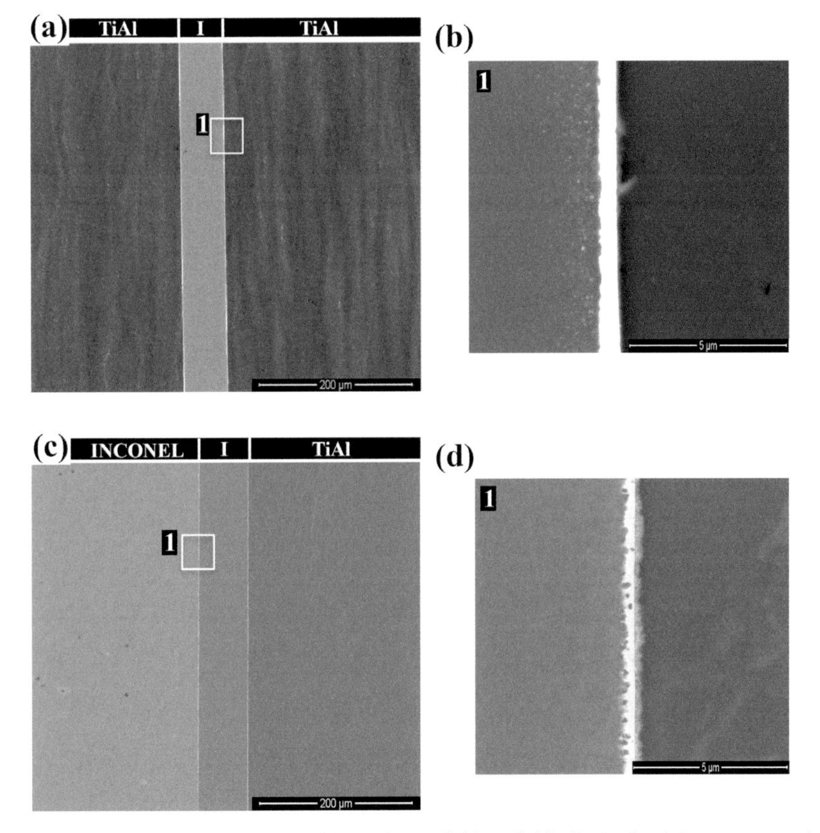

Figure 8.5 SEM images of (a) and (b) similar and (c) and (d) dissimilar joints processed at 700°C with NanoFoil®.

small particles rich in silver was observed; at the center of the interface, a large zone of NiAl grains formed due to the multilayer reaction. These interfaces were harder than those produced at room temperature. This higher value could be due to the nanometric grain size of the multilayers when the reaction occurs as a result of annealing with a slow heating rate. An increase in the bonding temperature to 800°C produced similar interfaces.

For dissimilar joints (γ-TiAl to Inconel 718), it was also possible to obtain successful joints with commercial Ni/Al multilayer foils at a high temperature. A sound joint, without pores or cracks, was formed at 700°C for 60 min under a pressure of 5 MPa. The interface presented six different zones with layers rich in silver close to the two base materials. The interface was harder than the base materials, with the highest value being measured at its center, as expected.

The strength of γ-TiAl joints produced with NanoFoil® by ignition with an electrical discharge and by annealing at 700 and 800°C was evaluated by shear tests. These results revealed that the joints produced with NanoFoil® exhibited a very low strength (Simões et al. 2014). The interfaces produced by annealing revealed a slightly higher value of shear strength, but this still remains very low when compared with the values of the interfaces produced only with multilayers. The fracture surfaces are smooth, showing several cracks in the NanoFoil®, which is an evidence of brittle behaviour. The fracture occurs mainly between the silver-rich superficial layer of the NanoFoil® and the base material. The silver-rich layer does not have a positive effect on the mechanical properties, contrary to expectations. The weak adhesion of the NanoFoil® to the base materials seems to be the factor that limits the strength of these joints. Some authors have solved the problems of adhesion by coating the base materials with brazing alloys (Duckham et al. 2004b), or with nickel and gold films (Wang et al. 2003, 2004, Duckham et al. 2004a), prior to joining. Although it is necessary to further improve the bonding process to produce an interface without cracks and with higher mechanical properties, the application of NanoFoil® to the joining of γ-TiAl alloys is a very interesting process, especially for structural applications. The use of reactive multilayers with different coatings may be one option for obtaining better results.

8.4.2 Joining of γ-TiAl Alloys Coated with Reactive Multilayers Combined with Brazing Alloys

The joining of γ-TiAl alloys can also be performed by coating the base materials with multilayers and placing a brazing alloy between the two coated mating surfaces. The use of this complex interlayer improves the adhesion of the base material/multilayer interface and combines the benefits of solid-state diffusion bonding with diffusion brazing.

This approach was tested for the joining of γ-TiAl alloys. Base materials coated with Ni/Al multilayers, with bilayer thicknesses of 5 and 14 nm and a total film thickness ranging from 2.0 to 2.7 μm, were combined with a freestanding TiNi 67 (Ti-26.7Ni at.%) brazing alloy 50 μm thick. The joints were performed at 800°C for 60 min under a pressure of 5 MPa. SEM images of the interfaces can be seen in Figure 8.6. EDS chemical compositions and phase identification of the zones marked in Figure 8.6 are listed in Table 8.3. A relevant aspect of the conditions required for producing these joints is the fact that the temperature is lower than the melting point of the brazing alloy, which supports the positive influence of the use of multilayers on reducing the processing conditions.

As can be seen in Figure 8.6, the interface exhibited some layers and several phases, resulting in chemical and structural discontinuities. Besides, although the apparent formation of sound joints, pores and cracks were

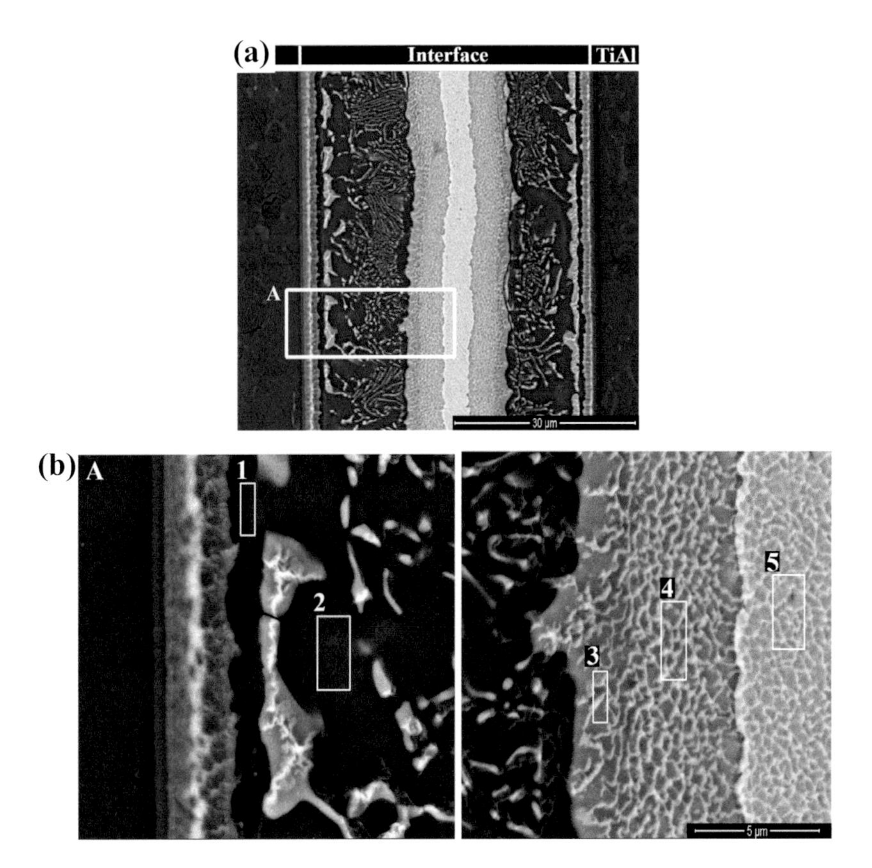

Figure 8.6 SEM images of the interface produced at 800°C with the combination of Ni/Al multilayers and TiNi 67 brazing alloy.

Table 8.3 EDS chemical composition (at.%) and phase identification of zones marked in Figure 8.6.

Zones	EDS composition (at.%)						Phases
	Al	Ti	Ni	W	Nb	Cr	
γ-TiAl alloy	47.9	46.9	—	0.3	3.1	1.8	γ-TiAl
1	23.9	73.9	2.2	—	—	—	α_2-Ti_3Al
2	4.1	94.3	1.6	—	—	—	α(Ti)
3	1.3	68.3	30.4	—	—	—	$NiTi_2$
4	0.9	47.9	51.2	—	—	—	B2-NiTi
5	1.0	24.1	74.9	—	—	—	Ni_3Ti

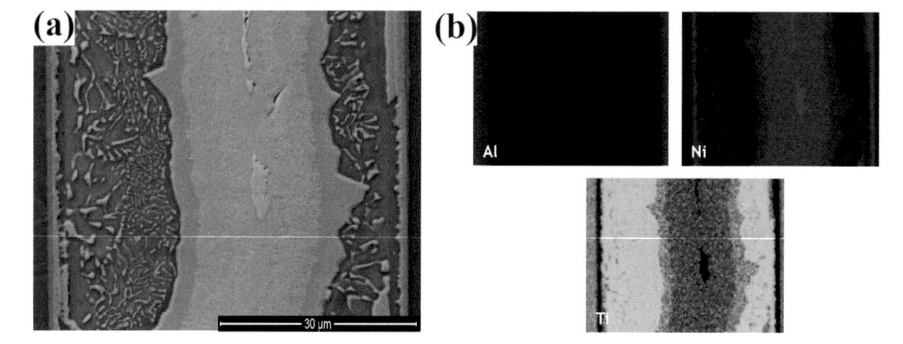

Figure 8.7 (a) SEM image of the interface processed at 800°C and (b) elemental maps across the interface of aluminum, nickel and titanium.

observed in localised areas. Figure 8.7 shows the EDS elemental maps of aluminum, nickel and titanium through the entire interface. Drawing on these results, it can be noted that the diffusion of elements through the interface during the joining process is responsible for the formation of the various layers.

The analysis of the EDS results with Ni-Ti, Ni-Al and Al-Ni-Ti phase diagrams (Baker and Okamoto 1992, Huneau et al. 1999) has enabled the identification of the various phases formed; from the γ-TiAl alloy to the center of the interface have been identified: α_2-Ti_3Al, α(Ti), $NiTi_2$, B2-NiTi and Ni_3Ti. The two thin layers close to the base material are rich in nickel and aluminum and are the result of the diffusion between the multilayers and the base material; similar phases were detected in the joining of γ-TiAl alloys using only Ni/Al reactive multilayers. Interdiffusion between the brazing alloy and the multilayers promoted the formation of α_2-Ti_3Al. The center of the interface is characterised by the presence of α(Ti), $NiTi_2$, B2-NiTi and Ni_3Ti resulting from the diffusion between the titanium and nickel of the brazing alloy. α(Ti) and $NiTi_2$ are the result of the eutectoid reaction that occurs at 765°C, through which β(Ti) is transformed to these two phases. The layers formed by the three intermetallic phases ($NiTi_2$, B2-NiTi and Ni_3Ti) remain during cooling. However, in these phases the reduction of nickel solubility with decreasing temperature (clearly visible for the NiTi phase in the Ti-Ni diagram (Baker and Okamoto 1992)) led to the formation of Ni-rich precipitates, which are the bright areas observed in SEM images of these intermetallics. The presence of brittle phases, such as α_2-Ti_3Al and Ni_3Ti, compromises the mechanical properties of the joints and may also be related to some pores and cracks observed in the layers formed by these phases. The shear strength of these joints is very low; for instance, the joints produced with a 14 nm bilayer thickness at 800°C, under a pressure of 5 MPa for 60 min, have shear strength of 36 ± 1 MPa. The analysis of the fracture surface of this sample shows a very flat surface, indicating poor bonding between the mating surfaces. Figure 8.8 shows the 3D map image of this fracture surface, showing the described microstructural characteristic. The fracture occurred along the α_2-Ti_3Al layer of the interface. Thus, the insertion of a TiNi 67 brazing alloy between two γ-TiAl samples coated with Ni/Al multilayers drastically reduced the strength of the joints when compared with joints bonded only with the multilayers.

To sum up, the joining of γ-TiAl base materials coated with Ni/Al reactive multilayers and combined with the TiNi 67 brazing alloy is a process that could allow the joining of complex shapes and larger components

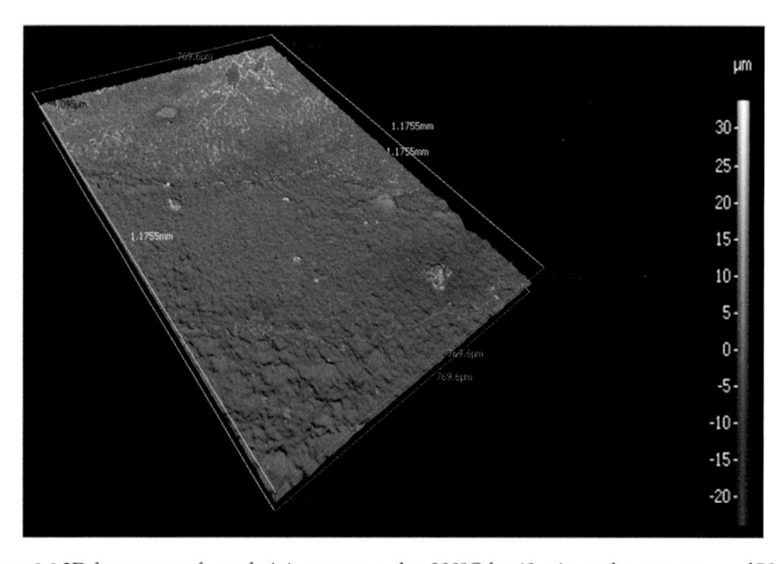

Figure 8.8 3D fracture surface of a joint processed at 800°C for 60 min under a pressure of 5 MPa.

in less stringent processing conditions than brazing. By comparing this process with that of brazing, using the same TiNi 67 alloy (Guedes et al. 2004, 2006), it was found that joining is feasible at a significantly lower temperature (800°C by this process and 1,050°C by brazing). However, some pores and cracks were observed in the interface that can be associated with the formation of very brittle phases, such as α_2-Ti_3Al and Ni_3Ti, caused by the reaction of the elements of the three constituents (base material/multilayers/brazing alloy).

The considerable thickness of the brazing alloy in comparison with the thickness of the reactive multilayers may be a determining factor in the failure of this system, since the thickness of the multilayer used is too thin to release the heat required to completely melt the brazing alloy during joining. The use of thicker multilayers and/or thinner brazing alloys may lead to better results. To achieve these, bonding experiments were performed using thin foils of titanium and nickel combined with Ni/Al multilayers deposited by sputtering onto the γ-TiAl alloys. A nickel foil, with a thickness of 1 μm, surrounded by titanium foils, with a thickness of 5 μm, were used as an interlayer. These foils were selected to achieve a composition close to the TiNi 67 brazing alloy while reducing the thickness of the interlayer significantly (from 50 to 11 μm). Ni/Al multilayers with bilayer thicknesses of 5, 14 and 30 nm and a total film thickness ranging from 2.0 to 2.5 μm were tested.

The smallest thickness of this system promotes better use of the heat released by the multilayers during joining, promoting localised heating of the thin titanium and nickel foils. Figure 8.9 shows the SEM images of the interfaces produced at 800 and 900°C for 30 min using these foils as interlayer. Microstructural analysis shows that interfaces well bonded and free of defects were produced. At 800°C only the Ni/Al reactive multilayers with smaller bilayer thicknesses of 14 and 5 nm formed an interface with apparent soundness.

The solid-state diffusion across the interface promotes the formation of six zones from the center of the interface to the base material, which are clearly identified in Figure 8.9. The different grey tones in the backscattered electron (BSE) images show the variation in composition along the interface. It should be noted that the microstructure of the γ-TiAl base material was unaffected by the joining procedure. Raising the bonding temperature from 800 to 900°C promotes an increase in total thickness of the interface; for joints using a bilayer thickness of 14 nm, the interface thickness changes from 18.1 to 21.7 μm with this temperature rise. This increase in total thickness of the interface is mainly caused by the layers formed between the multilayers and the titanium foils and was apparently the result of a more intense diffusion of aluminum.

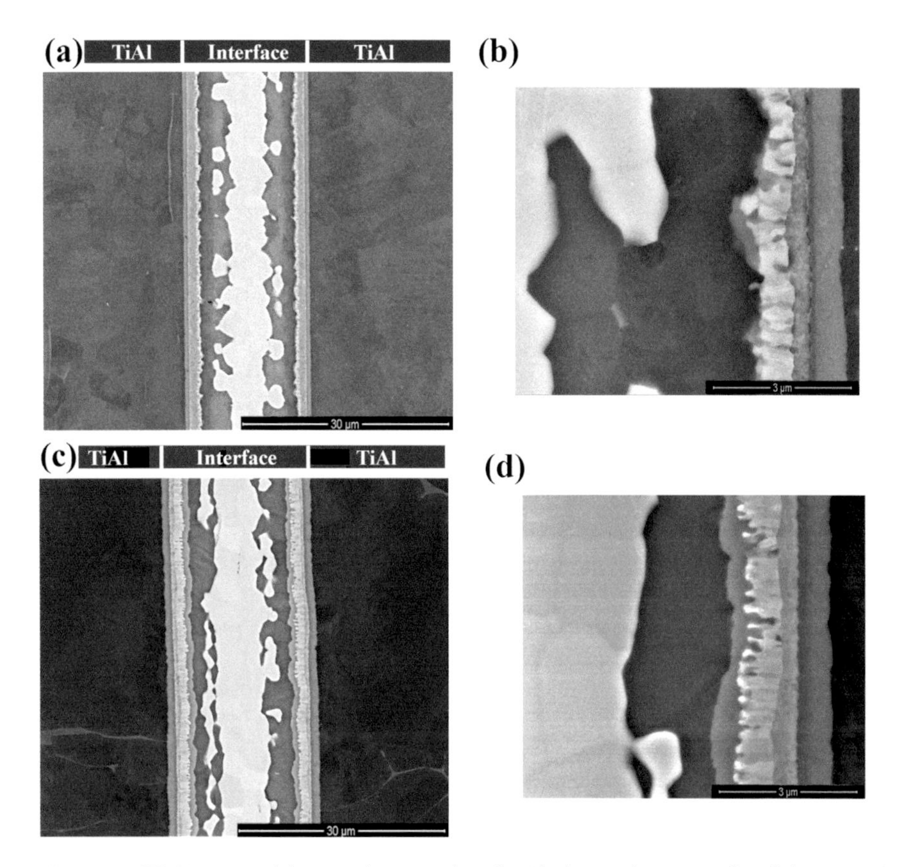

Figure 8.9 SEM images of the interfaces produced with the combination of multilayers and titanium and nickel thin foils obtained at: (a) and (b) 800ºC and (c) and (d) 900ºC.

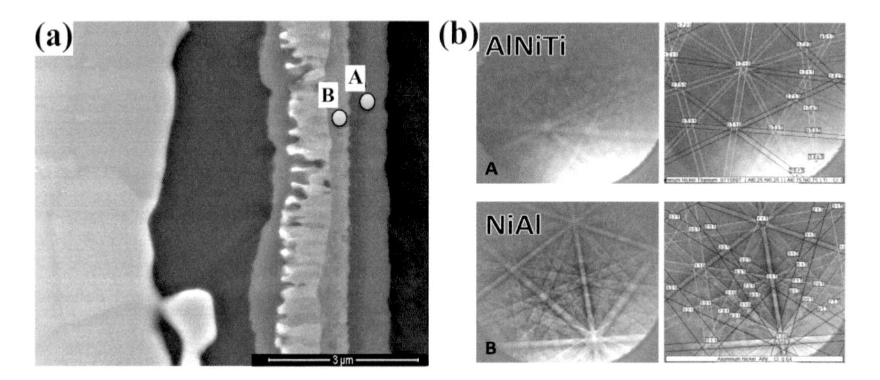

Figure 8.10 (a) SEM image and (b) EBSD Kikuchi pattern of the two layers close to the γ-TiAl alloy for joints processed with the combination of Ni/Al multilayers and nickel and titanium thin foils indexed as (A) AlNiTi and (B) NiAl.

The reaction products formed at the interfaces were characterised using a combination of EDS and Electron Backscatter Diffraction (EBSD) pattern analyses. EBSD patterns are particularly important for the thinner layers where the layer thickness is smaller than the volume of interaction of EDS measurements. Figure 8.10 shows two examples of EBSD pattern indexation of the thin layers closest to the γ-TiAl base material. The distribution of the elements across the interface can be observed in the EDS elemental maps of Figure 8.11. Based on these results, an evolution of the phases formed at the interface can be drawn for each joining temperature. At 800°C, interdiffusion between γ-TiAl (base material) and Ni/Al (multilayers) led to the formation of an AlNiTi layer; the Ni/Al multilayers were transformed into NiAl during the heating cycle. Moving towards the center of the interface, the region after the NiAl grains is formed by phases enriched in titanium, namely $AlNi_2Ti$ and α_2-Ti_3Al, in that order. These two phases resulted from the diffusion between the multilayers and titanium foils. Despite intense diffusion of aluminum, there remains a large untransformed titanium region (the dark region in the center of the interface). The bright layer observed in the central region has a composition similar to the $NiTi_2$ phase. This layer is formed due to interdiffusion of titanium and nickel from the two titanium foils and the central nickel foil.

At 900°C, the diffusion is more intense and the aluminum content at the interface is higher than that measured after bonding at 800°C. The intense diffusion of aluminum led to the formation of a new Al-rich phase (Al_2NiTi) in the layer adjacent to the base material; this region contains grains of two different phases, AlNiTi and Al_2NiTi. The interdiffusion of the multilayers and the titanium foil led to a phase evolution similar to that observed at 800°C. However, a continuous AlNiTi layer, which was not detected in this region at 800°C, developed between the $AlNi_2Ti$ and α_2-Ti_3Al layers. In addition, the α_2-Ti_3Al layer formed at 900°C is thicker than the one formed at 800°C and the titanium phase is not present. At the center of the interface,

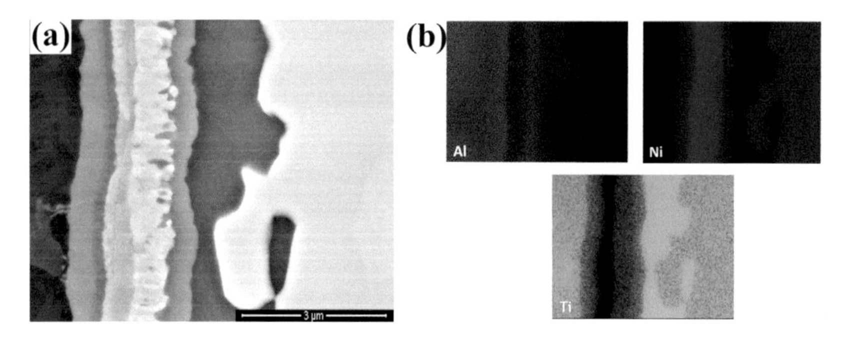

Figure 8.11 (a) SEM image of the interface processed at 800°C and (b) elemental maps across the interface of aluminum, nickel and titanium.

the $NiTi_2$ layer is again observed; at this temperature, this phase contains dissolved 9.3 at.% aluminum.

The hardness and shear strength were measured to evaluate the mechanical properties of the diffusion bonds. The reaction layers are harder than the γ-TiAl base material (10 GPa), except the α(Ti) regions at 800°C (9 GPa) or α_2-Ti_3Al regions at 900°C (9 GPa) (see Figure 8.12). For both temperatures, the highest hardness values are attained for the layer closest to the base material (14 GPa at 800°C and 19 GPa at 900°C). The hardest intermetallic phases are the AlxNixTi intermetallics (x=1 or 2); AlNiTi+Al_2NiTi is the hardest zone, followed by the zones formed by the AlNiTi and $AlNi_2Ti$ phases. Some phases (AlNiTi, $NiTi_2$ and α_2-Ti_3Al) present different hardness values depending on bonding temperature, possibly as a result of composition and grain size differences. The increase in diffusivity in line with temperature rise, leading mainly to a general enrichment in aluminum across the interface, is responsible for a solid solution hardening effect, which is added to the formation of hard intermetallic phases.

The hardness variation across the interface associated with the formation of continuous layers of very hard intermetallics may be a problem for the performance of the joint. Actually, the shear strength of these joints is very low, and this combination of multilayer thin films with nickel and titanium foils reduces the shear strength of the joints when compared with those bonded only by multilayers; in fact, joints processed with the same Ni/Al multilayer (with a bilayer thickness of 14 nm) under similar conditions (900°C under a pressure of 5 MPa) exhibit shear strengths of 31 ± 1 MPa or 314 ± 40 MPa depending on whether they were or were not obtained using nickel and titanium foils. Despite this decrease in mechanical performance of the joint at room temperature, improvement was observed when comparing

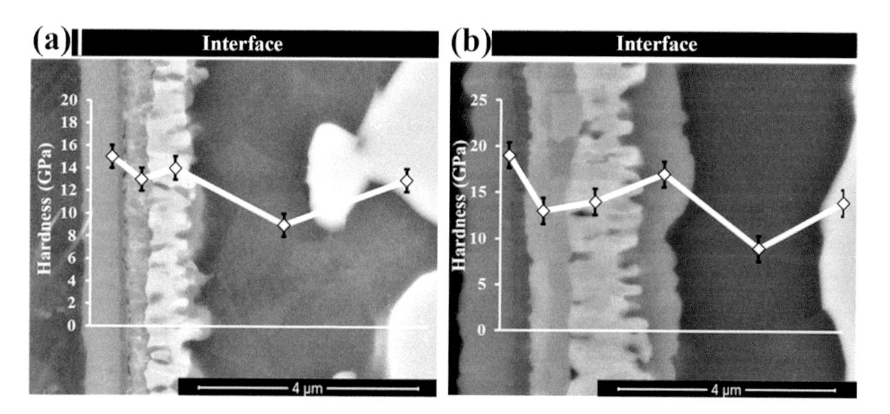

Figure 8.12 SEM images of the interface processed at (a) 800°C and (b) 900°C with overlapping hardness profiles obtained by nanoindentation tests.

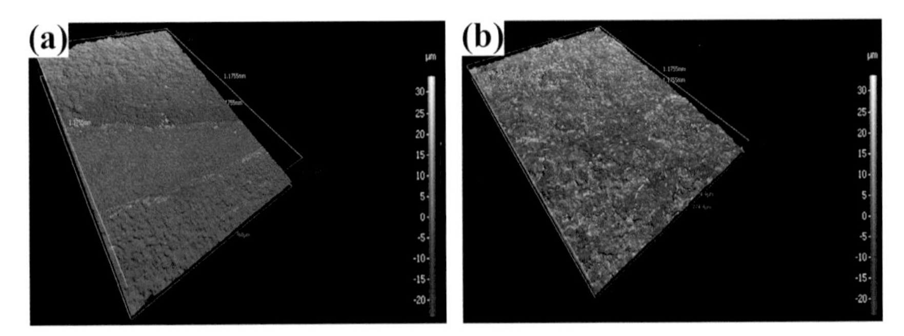

Figure 8.13 3D fracture surface of the joint processed for 60 min under a pressure of 5 MPa at (a) 800°C and (b) 900°C.

these bonds with those that use a combination of multilayers and thicker TiNi 67 brazing alloy.

The fracture surfaces of shear test specimens bonded using Ni/Al multilayers with 14 nm of bilayer thickness combined with titanium and nickel foils were examined by SEM. Figure 8.13 shows a 3D image map of the fracture surface of these diffusion bonds processed at 800 and 900°C. This fracture surface is very smooth, with small plateau areas that seem to traverse different zones of the interface. EDS and EBSD results revealed that the fracture occurs mainly through the $NiTi_2$, AlNiTi and $AlNi_2Ti$ phases. However, it is also possible to identify α_2-Ti_3Al and Al_2NiTi phases in localised areas. These observations explain the low strength of the joints. The fracture occurred entirely in the interface and mainly between the layers that have higher hardness values. A unique feature of the central area of the interface is the large grain size (close to 5 μm) of $NiTi_2$ grains, while the other zones have a smaller grain size, less than 1 μm.

The combination of Ni/Al multilayers and titanium and nickel foils in joining γ-TiAl alloys has been used in order to enhance contact between the surfaces to be joined. Despite the low strength of these joints, replacement of the TiNi 67 brazing alloy by titanium and nickel foils led to an improvement in the quality of the joint.

8.5 Concluding Remarks

This chapter has described a different approach to improving the joining processes, especially for γ-TiAl alloys. This approach consists in the use of a combination of reactive multilayers and brazing alloys to associate the benefits of heat released by the multilayers, and their nanometric character, with the capacity of brazing alloys to fill the gaps in the mating surfaces. The reactive multilayers and brazing alloys can be used as freestanding films, but deposition of one of these materials onto the base material, or of one coating onto the other, reduces the problems in the adhesion of an interface (interface base material/multilayers or base material/brazing alloy or multilayers/brazing alloy). Ni/Al multilayers coated with brazing alloys have already been commercialised (NanoFoil® made by Indium Corporation) for use as an interlayer in order to reduce the bonding conditions, particularly the bonding temperature. In some systems, the use of the NanoFoil® allows the joining to be processed at room temperature. However, the successful application of NanoFoil® in joining processes is strongly dependent on the materials to be joined.

The joining of γ-TiAl alloys can be performed at room temperature using NanoFoil®. The main problem consists in the formation of cracks perpendicular to the interface, which compromises the joint performance. This occurs either by the multilayers contracting during the reaction or by the fracture of the NanoFoil® during the application of pressure prior to ignition of the reaction, as a result of residual stresses and brittleness of these nanolayers. To prevent these cracks, leading to a better bond, joining of γ-TiAl alloys can be performed at 700°C using the NanoFoil®. The interfaces are free from defects, but the shear strength is very low. The fracture occurs mainly through the silver-rich superficial layer of the NanoFoil® and the base material. The silver-rich layer, which acts as a brazing alloy, does not have a positive effect, contrary to expectations, and weak adhesion to the base material compromises the mechanical performance.

The combination of Ni/Al multilayers and TiNi 67 brazing alloy or titanium and nickel foils (with thickness selected to achieve a composition close to that of a commercial TiNi 67 brazing alloy) in the joining of γ-TiAl alloys has proved effective for improving contact between the surfaces to be joined. The titanium and nickel foils have led to better quality of the interface. By using this combination, it is also possible to reduce the bonding temperature to 800°C, which is below the melting temperature of the brazing alloy. However, for both combinations tested, the shear strength is lower than when only the base material coated by Ni/Al multilayers is used.

Summing up, the best approach to joining γ-TiAl alloys seems to be the use of reactive multilayers coating the mating surfaces. However, for applications where the mechanical performance is not critical, a combination of multilayers and brazing systems can be used, with the advantages of reduced surface preparation requirements.

Keywords: Bilayer thickness; brazing alloys; hardness; joining processes; mechanical properties; microstructure; NanoFoil®; phase formation; reaction; reactive multilayers.

8.6 References

Baker, H. and H. Okamoto [eds.]. ASM Handbook, Volume 03 - Alloy Phase Diagrams. 1992. ASM International, USA.

Bartout, D. and J. Wilden. 2012. Combined scale effects for effective brazing at low temperatures. MATEC Web Conf. 1: 00004.1-3.

Duckham, A., M. Brown, E. Besnoin, D. Van Heerden, O.M. Knio and T.P. Weihs. 2004a. Metallic bonding of ceramic armor using reactive multilayer foils. Ceram. Eng. Sci. Proc. 25: 597–603.

Duckham, A., S.J. Spey, J. Wang, M.E. Reiss, T.P. Weih, E. Besnoin et al. 2004b. Reactive nanostructured foil used as a heat source for joining titanium. J. Appl. Phy. 96: 2336–2342.

Guedes, A., A.M.P. Pinto, M.F. Vieira and F. Viana. 2004. Joining Ti-47Al-2Cr-2Nb with a Ti-Ni braze alloy. Mater. Sci. Forum 455-456: 880–884.

Guedes, A., A.M.P. Pinto, M.F. Vieira and F. Viana. 2006. Assessing the influence of heat treatments on γ-TiAl joints. Mater. Sci. Forum 514-516: 1333–1337.

Huneau, B., P. Rogl, K. Zeng, R. Schmid-Fetzer, M. Bohn and J. Bauer. 1999. The ternary system Al-Ni-Ti. Part I: Isothermal section at 900°C; Experimental investigation and thermodynamic calculation. Intermetallics 7: 1337–1345.

Indium Corporation. 1996–2016. NanoFoil made by Indium Corporation. https://www.indium.com/nanofoil/.

Kim, J.S., T. LaGrange, B.W. Reed, M.L. Taheri, M.R. Armstrong, W.E. King et al. 2008. Imaging of transient structures using nanosecond *in situ* TEM. Science 321: 1472–1475.

Qiu, X. and J. Wang. 2008. Bonding silicon wafers with reactive multilayer foils. Sensor. Actuator. 141: 476–481.

Simões, S., F. Viana, A.S. Ramos, M.T. Vieira and M.F. Vieira. 2011. Anisothermal solid-state reactions of Ni/Al nanometric multilayers. Intermetallics 19: 350–356.

Simões, S., F. Viana and M.F. Vieira. 2014. Reactive commercial Ni/Al nanolayers for joining lightweight alloys. J. Mater. Eng. Perform. 23: 1536–1543.

Simões, S., A.S. Ramos, F. Viana, O. Emadinia, M.T. Vieira and M.F. Vieira. 2016. Ni/Al multilayers produced by accumulative roll bonding and sputtering. J. Mater. Eng. Perform. 25: 4394–4401.

Sun, X. 2010. Joining lightweight materials using reactive nanofoils. pp. 289–306. *In*: X. Sun. (ed.). Failure Mechanisms of Advanced Welding Processes. CRC Press, Woodhead Publishing Limited, Cambridge, England, UK.

Trenkle, J.C., L.J. Koerner, M.W. Tate, N. Walker, S.M. Gruner, T.P. Weihs et al. 2010. Time-resolved x-ray microdiffraction studies of phase transformations during rapidly propagating reactions in Al/Ni and Zr/Ni multilayers foils. J. Appl. Phys. 107: 113511-1-12.

Wang, J., E. Besnoin, A. Duckham, S.J. Spey, M.E. Reiss, O.M. Knio et al. 2003. Room-temperature soldering with nanostructured foils. Appl. Phy. Lett. 83: 3987–3989.

Wang, J., E. Besnoin, O.M. Knio and T.P. Weihs. 2004. Investigating the effect of applied pressure on reactive multilayer foil joining. Acta Mater. 52: 5265–5274.

Index